LET'S WORK SAFELY!

English Language Skills For Safety In The Workplace

Linda Mrowicki

Illustrated by
Sally Richardson

LINMORE PUBLISHING, INC. P.O. BOX 1545 PALATINE, IL

Linmore Publishing Inc.
P.O. Box 1545
Palatine, IL 60078

First printing, 1984 Printed in the United States of America
Second printing, 1986

LET'S WORK SAFELY!
Student's Book ISBN 0-916591-00-X
Teacher's Book ISBN 0-916591-01-8

ACKNOWLEDGEMENTS

The safety signs and posters found in the exercise "Read the Safety Signs and Posters" are reproduced through the courtesy of the National Safety Council. Safety signs, stickers, posters, forms, and other materials are available from the National Safety Council. For information, contact the National Safety Council, 444 N. Michigan Ave., Chicago, IL 60611.

The statistics for the numbers of work-related accidents, injuries, and deaths are from the National Safety Council publications *Accident Facts* and *The Disablers–Know Your Four Worst Enemies.*

Several people provided valuable assistance in the development of this book. Their comments and suggestions for the content, methodology, and activities are greatly appreciated.

The following people were useful resources for the safety content: **Joseph Eckerman**, Job Placement Counselor at the International Institute of Akron, Ohio and Adult Vocational Machine Instructor at the Akron Board of Education; **Daniel Koch**, Vice-President, Morris-Kurtzon, Inc., and **Gregory Morris**, Director of Field Engineering, Chicago Power and Process and Electronics Technology Instructor at the College of Lake County.

The following people provided input for the language content and methodology: **Kamla Devi Koch**, Independent Consultant; **Catherine Robinson**, Employment Class Coordinator at the International Institute of St. Paul, Minnesota; **Jenise Rowecamp**, Consultant for the Minnesota Department of Education; **Arlene Ruttenberg**, Curriculum Specialist at the ICMC Philippine Refugee Processing Center; **Dennis Terdy**, Director of the Illinois ESL/Adult Education Service Center, and **Susan Thompson**, ESL Instructor at the Elgin YWCA Refugee Program. Ms. Thompson participated in the field-testing of the materials and offered many valuable suggestions. Special thanks goes to all the students who used the materials and provided helpful feedback.

I would also like to thank **Mr. Melvin Boruszak** (Mr. "B") of S.C.O.R.E., for all his assistance in getting this book printed.

Especially important to me was the interest, enthusiasm, and support of my family in this endeavor– **Daniel C. Jackson**, **Genevieve Mrowicki**, **James Mrowicki**, and **Joseph Mrowicki**.

TABLE OF CONTENTS

LET'S WORK SAFELY COURSE OUTLINE

Lesson	Competency	Grammatical Structure	Vocabulary	Safety Topic
1. Safety Is Important!	Identify numbers of accidents and deaths.	Simple present Simple past	Numbers Body parts	Importance of safety
2. Everyone Is Concerned About Safety	Become aware that safety is important.	Simple present	Occupations	Accidents can happen in any job
3. Proper Dress	Comprehend & give instructions about proper dress.	Imperatives It/you could get...	Clothing Hazards	Proper & improper dress
4. Types Of Safety Clothing	Comprehend & give instructions about safety clothing.	You have to... Simple present, 3rd person	Safety clothing Body parts	Importance of wearing safety clothing
5. Getting Safety Clothing	Request safety clothing.	I need a/some...	Safety clothing	Sources of safety clothing
6. Insisting!	Insist on getting safety clothing.	I need a/some... There is/are...	Safety clothing	Acceptance of responsi-bility for safety
7. Proper Behavior	Comprehend & give instructions for proper behavior.	Don't... Stop ___ ing	Types of unsafe behavior	Safety rules for behavior
8. Lift! Move! Stack!	Follow & give instructions for lifting, moving, & stacking.	Imperatives You should...	Mechanical aids	Proper lifting, moving & stacking
9. Dangerous Materials	Ask about & explain safety procedures. Read instructions.	Imperatives	Flammable, poisonous, combustible	Using & handling dangerous materials
10. Cautions	Comprehend & express cautions.	Don't...You can't...You'd better not...before __ing	Hand tools, power tools, machines	Safety rules for using tools
11. Is It Safe?	Inquire about unsafe working conditions.	There is/are Questions	Unsafe conditions	Unsafe conditions
12. Reporting An Unsafe Condition	Report an unsafe condition.	Be, present tense I'd like to report...	Working conditions	Responsibility for reporting hazards
13. Suggestions	Make suggestions for improving working conditions.	I think... If..., it would...	Working conditions	Responsibility for making suggestions
14. Warnings	Warn others of dangers.	Be, present tense Present progressive	Duck! Look out! etc.	Potential dangers
15. Fire!	Comprehend & express warnings about fire. Report fires.	Imperative	Fire hose, fire extinguisher, etc.	Safety rules for fire prevention
16. A Minor Accident	Give & accept advice politely.	Simple past You should...	Injuries – cuts, bruises, etc.	Taking immediate care of injuries
17. Reporting An Injury	Report accidents & injuries.	Simple past	Types of injuries	Necessity of reporting injuries
18. How Did It Happen?	Identify events preceding an accident.	Past progressive Simple past	Hazards–spills, broken handrails, etc.	Reporting procedures & forms
19. More Suggestions	Make suggestions for preventing accidents.	I think... It should...	Hazards	Responsibility for accident prevention

INTRODUCTION

LESSON ONE: Safety Is Important!

You will learn about the accidents and injuries that happen each year.

LESSON TWO: Everyone Is Concerned About Safety

You will learn that safety is important in every job.

LESSON ONE: SAFETY IS IMPORTANT!

Safety in the workplace is very important. Workers have to be careful. There are many accidents in the workplace. In 1990, 1,800,000 workers were hurt. 10,500 workers died on the job. In 1989, 1,700,000 workers were hurt. 10,700 people died on the job.

1,800,000 workers were hurt in 1988. 11,000 workers died on the job. Every year many people are hurt and many people die. WORKERS HAVE TO BE CAREFUL!

CHECK YOUR UNDERSTANDING.

1. How many workers were hurt in 1990? ____________________
2. How many workers were hurt in 1989? ____________________
3. How many workers were hurt in 1988? ____________________
4. How many workers died in 1990? ____________________
5. How many workers died in 1989? ____________________
6. How many workers died in 1988? ____________________
7. Is safety important? ____________________

There are many injuries every year. Here is an example of the kinds of injuries that happened in one year.

230,000 workers had head injuries.

190,000 workers had arm injuries.

150,000 workers had hand injuries.

320,000 workers had finger injuries.

610,000 workers had trunk injuries. (The trunk is the chest, stomach, hips, and back.)

270,000 workers had leg injuries.

100,000 workers had foot injuries.

40,000 workers had toe injuries.

CHECK YOUR UNDERSTANDING.

Look at the picture.
Write the names of the body parts and the numbers of injuries.

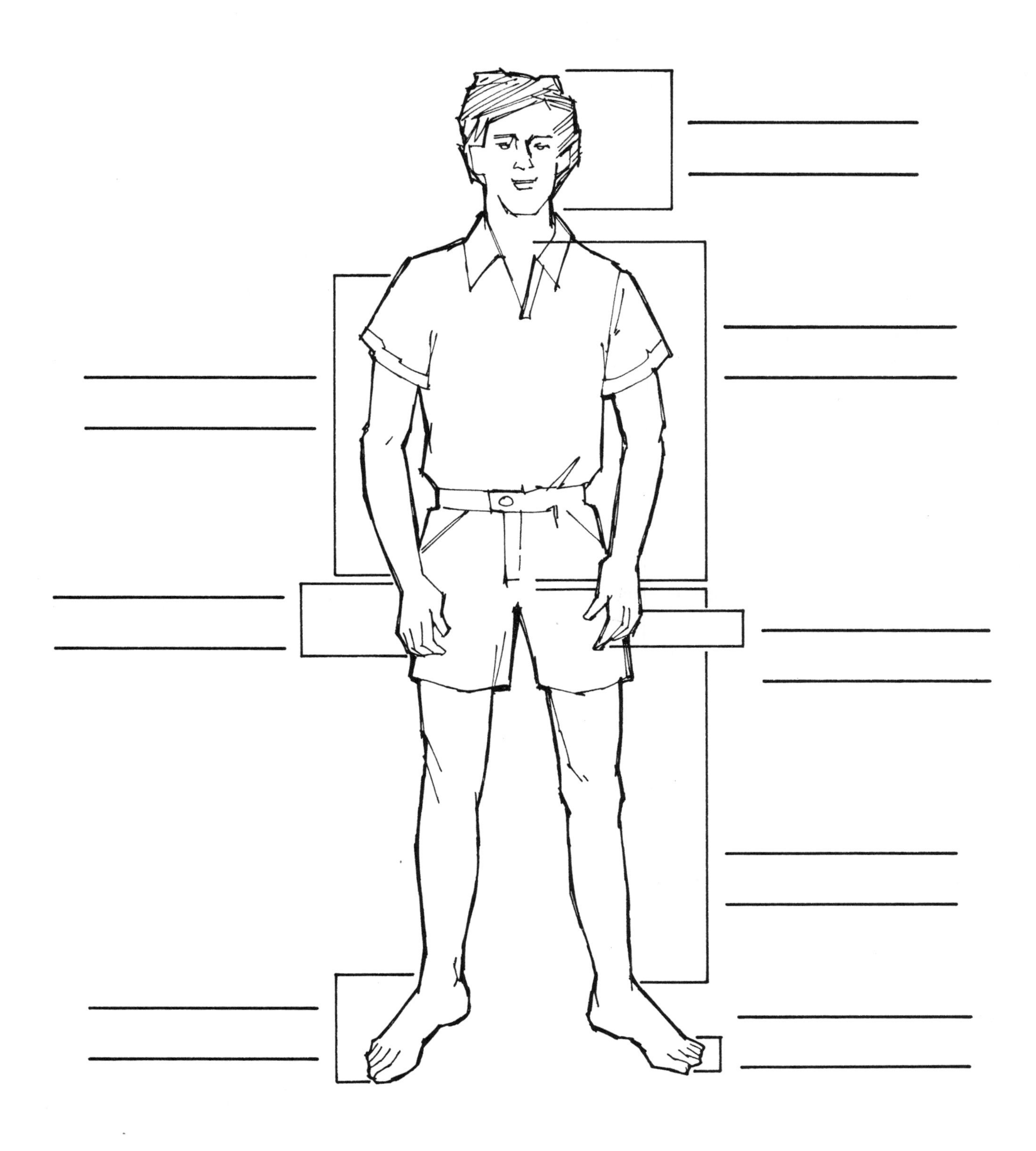

READ ABOUT SAFETY.

Everyone is concerned about safety. Employers do not want accidents. Supervisors check the workplace. Supervisors make safety rules for the workplace.

The government is also concerned about safety. The government makes laws about safety. Government inspectors check the workplace.

Many workers belong to unions. Unions are also concerned about safety. They check the workplace also.

Workers are also concerned about safety. It is very important for YOU to be concerned about safety.

Accidents can also happen in shop class. Shop teachers are very concerned about safety. Students should also be concerned.

English is very important. In this book you will study English and learn about safety at the same time. You will be able to talk about safety. You will be able to read safety signs and fill out safety forms. You will read many safety rules and learn about accident prevention.

CHECK YOUR UNDERSTANDING.

1. Who is concerned about safety?

    ____

2. Have you ever had an accident at work? ____________________

    What happened? ____________________________________

    ____

3. Is safety important to you? ____________________

4. What will you learn in this book?

    ____

    ____

LESSON TWO: EVERYONE IS CONCERNED ABOUT SAFETY

There are many people in this book. The people are like you. They do not want accidents. They are concerned about safety.

This is Champa Mounivong. She wants to find a job. She wants to be an assembler.

CHECK YOUR UNDERSTANDING.

I. What kind of job does Champa want?

____

2. Do you think safety is important to an assembler? ____________________

Chun Lee and Anwar Azami are students at a vocational school.

Chun is studying machine tools. He wants to be a machine operator after he graduates. He works with machines in machine shop.

Anwar is studying carpentry. He wants to be a carpenter after he graduates. He works with many tools in carpentry shop.

CHECK YOUR UNDERSTANDING.

1. What is Chun studying? ______________________________

2. What is Anwar studying? ______________________________

3. Do you think safety is important in shop class? __________________

These women work in a large hotel. Marta Popovich is a housekeeper. She cleans rooms. Kiem Nguyen works in the hotel kitchen. She is a food preparer. She makes salads for the hotel restaurant.

CHECK YOUR UNDERSTANDING.

1. Where do Marta and Kiem work? ______________________________

   ____

2. What jobs do they have? ____________________________________

   ____

3. Do you think safety is important to them? _______________________

   ____

These people work in a cookie factory. The company makes cookies. Eighty people. work in this factory.

Rosa Cruz works part-time. She is a sorter. She sorts cookies.
She packs cookies in boxes.

Bee Vang is a forklift driver. He loads and unloads many trucks. He takes boxes from one part of the factory to another part.

Safety is important to Bee and Rosa. They are concerned about safety. They are also concerned about the safety of the customers. They want the customers to buy good food.

CHECK YOUR UNDERSTANDING.

1. Where do Bee and Rosa work?

 ____

2. What jobs do they have? ______________________________

 ____

3. Do you think safety is important to them? ______________________________

These people work in the same factory. The company makes lamps and lights. There are 120 workers. All the workers belong to a union.

Stefan Michalski, Luis Martinez, and Johanis Dawit work with machines. Stefan is a machine operator. Luis is a welder. Johanis has several different jobs. Sometimes he is a machine operator. Other times he is a maintenance man.

CHECK YOUR UNDERSTANDING.

1. Where do Stefan, Luis, and Johanis work? ______________________________

   ____

2. What jobs do they have? ______________________________

   ____

3. Do you think safety is important to them? ______________________________

Each year there are many accidents and injuries. An accident can happen to anyone. All the people in this book are concerned about safety. They do not want any accidents to happen.

Stefan, Luis, Johanis, and Chun think about safety. They know that about 220,000 workers are hurt by machines each year.

Kiem thinks about safety. She knows that 5% of all injuries happen to food service workers.

Marta thinks about safety. She knows that 2% of all injuries happen to cleaning workers.

Bee thinks about safety. He knows that 1% of all injuries happen to forklift drivers.

Chun and Anwar think about safety. They know that many accidents happen in shop class.

CHECK YOUR UNDERSTANDING.

1. How many workers are injured by machines each year? ____________________
2. Do housekeepers have accidents? ____________________
3. Is there a completely safe job? ____________________
4. Are you concerned about safety? ____________________

SAFETY CLOTHING

LESSON THREE: PROPER DRESS

You will learn the rules for proper dress.

LESSON FOUR: TYPES OF SAFETY CLOTHING

You will learn the names of safety clothing.

LESSON FIVE: GETTING SAFETY CLOTHING

You will learn to ask for safety clothing.

LESSON SIX: INSISTING!

You will learn to insist on safety clothing.

LESSON THREE: PROPER DRESS

LISTEN.

Champa has a new assembly job. Today is her first day. Her supervisor is giving her an orientation.

Supervisor:	You'll be working on this machine. We have safety rules for an assembler. I see you have long hair. You have to tie it back. You can't wear your hair loose.
Champa:	Why?
Supervisor:	It could get caught in the machine. That would really hurt!
Champa:	Yes, I understand.
Supervisor:	Also, do not wear any jewelry. No rings, necklaces, earrings, or anything. OK?
Champa:	Why?
Supervisor:	Jewelry can get caught in the machine, too.
Champa:	OK, no jewelry.
Supervisor:	Any questions?
Champa:	No, I understand. No jewelry and I have to wear my hair back.
Supervisor:	Right. We don't want any accidents.

CHECK YOUR UNDERSTANDING.

1. Champa is an assembler.	RIGHT	WRONG
2. Champa can have long hair.	RIGHT	WRONG
3. Champa has to tie her hair back.	RIGHT	WRONG
4. It is safe for Champa to wear a necklace.	RIGHT	WRONG
5. Jewelry can get caught in the machine.	RIGHT	WRONG

PRACTICE

Look at the pictures. What is unsafe? Circle the problem.

Practice the conversation with your partner.
Practice conversations using the pictures below.

Supervisor:	Don't wear a necklace.
Worker:	Why?
Supervisor:	You could get hurt. It could get caught in the machine.
Worker:	OK. No necklace.

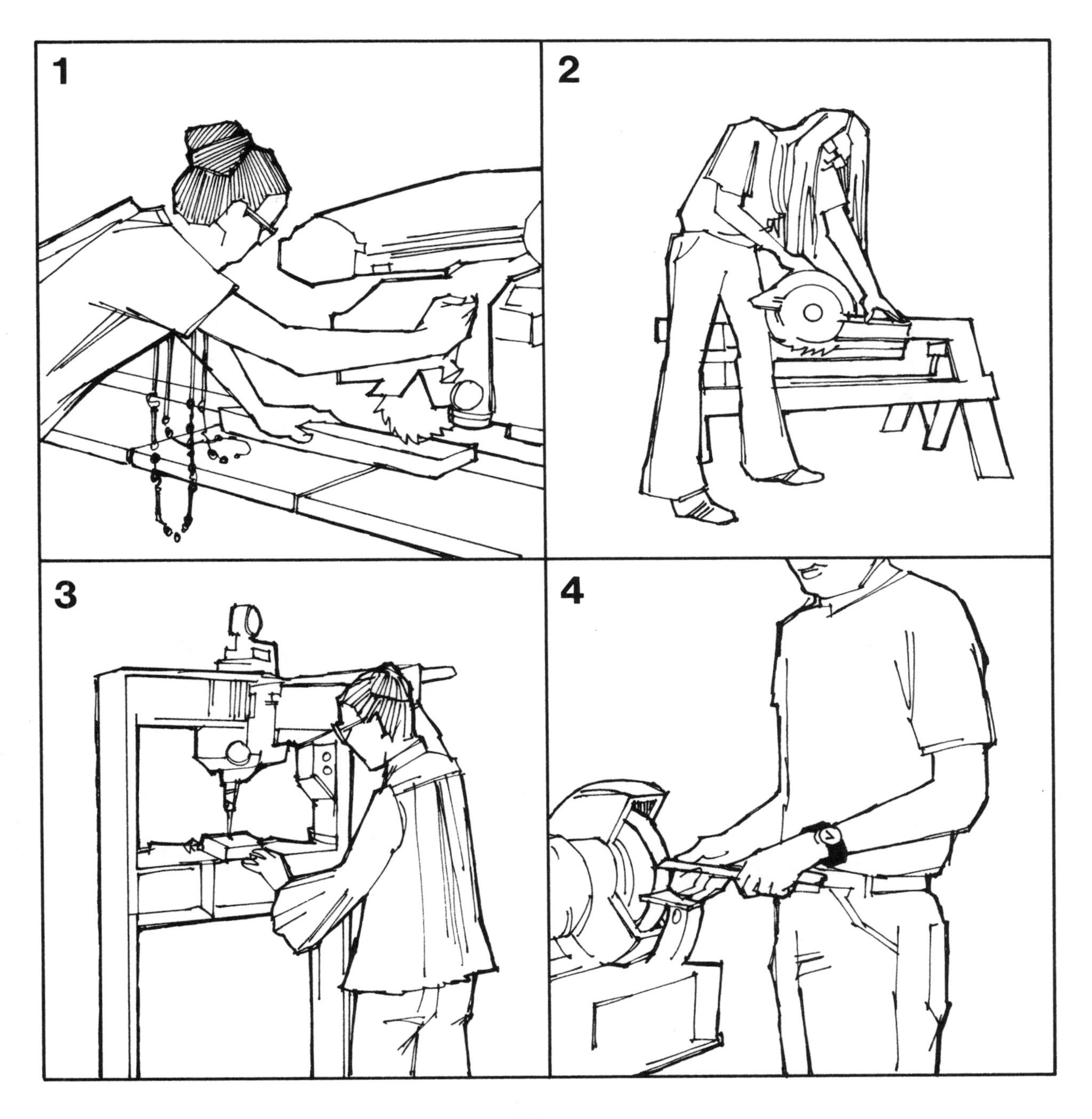

PRACTICE.

Look at the pictures. What is unsafe? Circle the problem.

Practice the conversation with your partner.
Practice conversations using the pictures below.

Supervisor:	Don't wear sandals.
Worker:	Why?
Supervisor:	You could get hurt. A box could fall on your foot.
Worker:	I understand. No sandals.

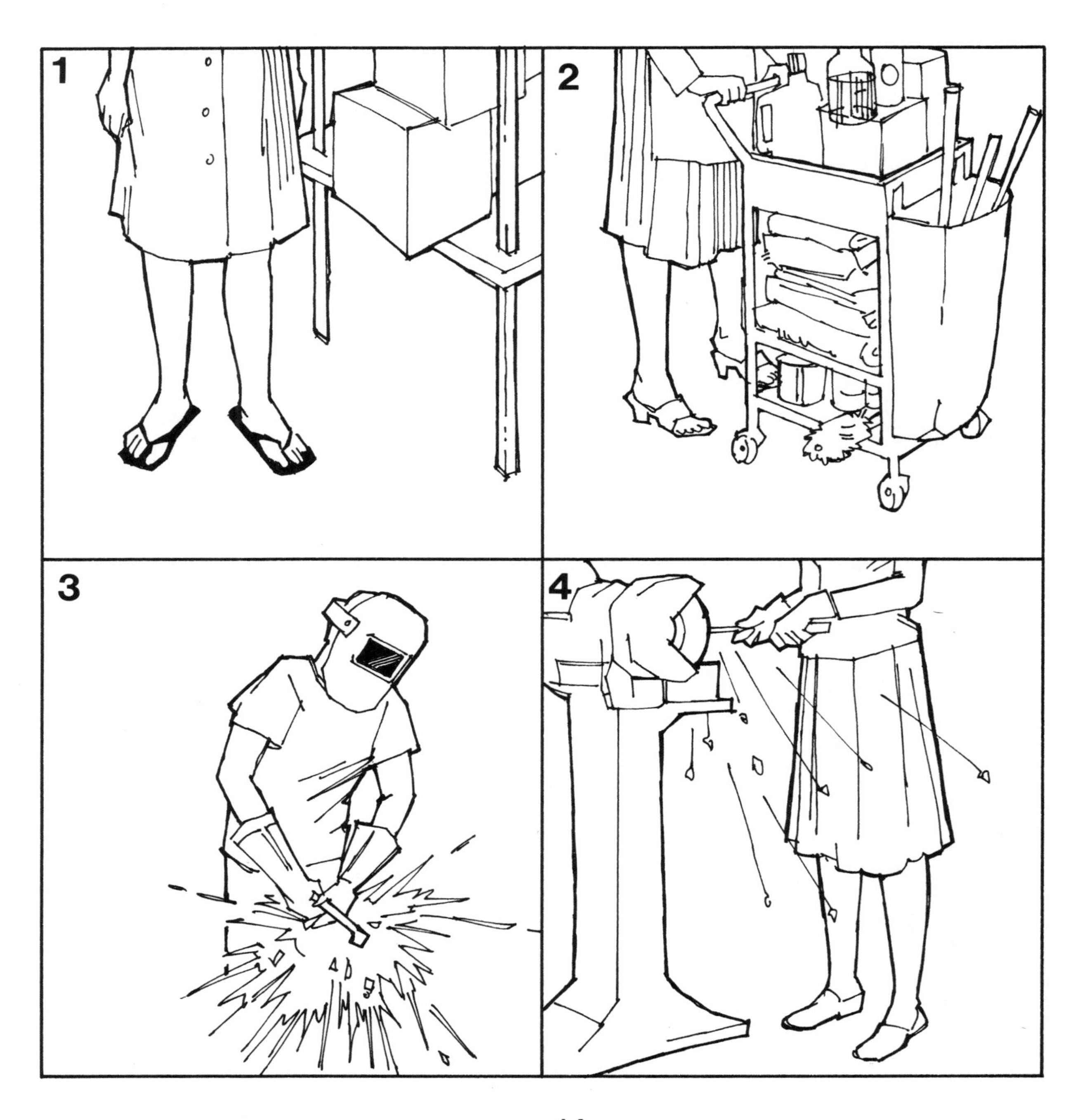

READ THE SAFETY POSTERS.
Answer the questions.

1. What shouldn't you wear?

____

____

Why not?

____

____

2. What should you wear?

____

____

Why?

____

____

READ ABOUT SAFETY.

Each workplace has rules about proper dress. It is very important to follow these rules. If you do not follow the rules, you could lose your job.

Here are some rules for Champa's workplace.

1. If you have long hair, tie it back.
2. Do NOT wear loose clothing.
3. Do NOT wear a tie or scarf if you are operating a machine.
4. Do NOT wear a watch or jewelry if you are operating a machine.
5. Wear good, strong shoes.
   Do NOT wear shoes with open toes or high heels.
   NO sandals or tennis shoes.

CHECK YOUR UNDERSTANDING.

1. Can a worker have long hair?

   ____

2. What does a worker have to do if he or she has long hair?

   ____

3. What are some examples of jewelry?

   ____

4. Can a worker wear tennis shoes?

   ____

5. What kind of shoes should a worker wear?

   ____

ROLE-PLAY.

Look at the pictures. What is unsafe? Circle the problem.

Read your role. Close your book. Role-play with your partner.

A

You are the supervisor. You tell the worker what he or she should not wear. You explain why proper dress is important.

B

You are the worker in the picture. Say OK if you understand the supervisor.

FOLLOW-UP.

Visit a workplace or shop class. Ask about the safety rules.
What are the rules for proper dress?

LESSON FOUR: TYPES OF SAFETY CLOTHING

LISTEN.

Stefan is a sheet metal cutter. He is starting a new job. His supervisor is giving him an orientation.

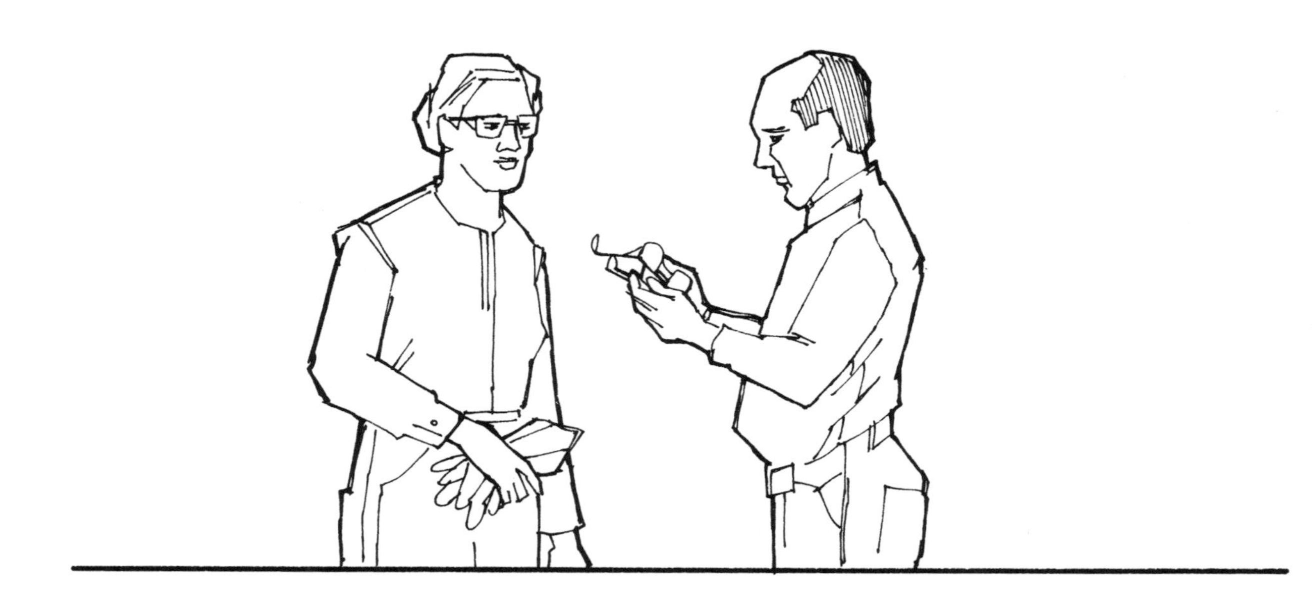

Supervisor:	Stefan, you'll be working on this machine. Now it is important that you wear your safety clothing.
Stefan:	What safety clothing?
Supervisor:	You have to wear safety glasses and gloves. They protect your eyes and hands.
Stefan:	Safety glasses and gloves. OK.

CHECK YOUR UNDERSTANDING

1.	Stefan is a metal cutter.	RIGHT	WRONG
2.	Stefan has to wear safety glasses.	RIGHT	WRONG
3.	Stefan has to wear gloves.	RIGHT	WRONG
4.	Stefan has to wear a hat.	RIGHT	WRONG
5.	Safety glasses protect a worker's hands.	RIGHT	WRONG

PRACTICE.

Practice the names of the safety clothing.

Write the names.

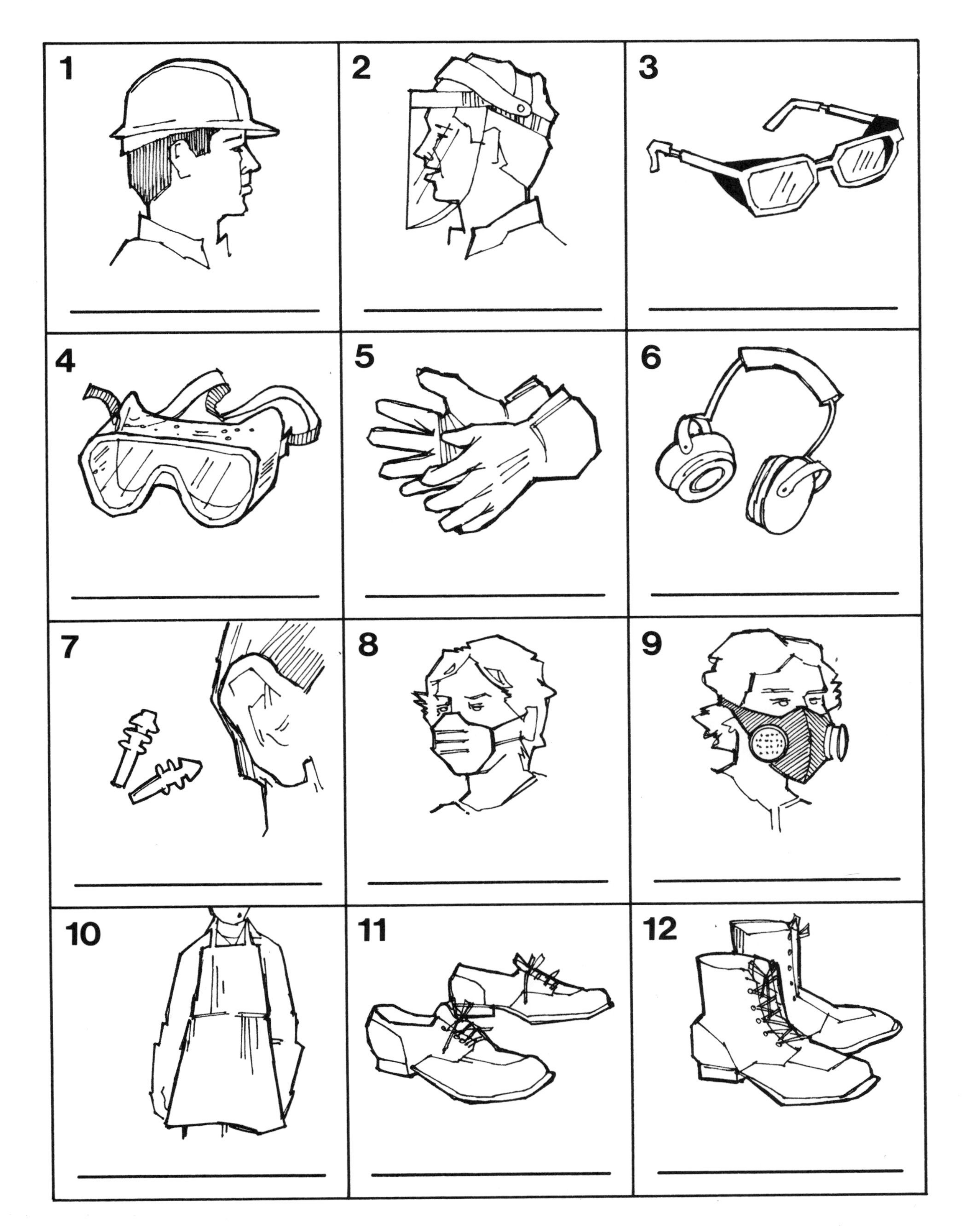

PRACTICE.

Practice these conversations with your partner.
Practice conversations using the pictures on page 19.

Supervisor:	You have to wear a hard hat.
Worker:	Why?
Supervisor:	It protects your head.
Worker:	A hard hat. OK.

Supervisor:	You have to wear ear plugs.
Worker:	Why?
Supervisor:	They protect your ears.
Worker:	Ear plugs. OK.

PRACTICE.

Some workers handle food. They wear special clothing.
They want to protect the customers.

Practice these names.
Write the names.

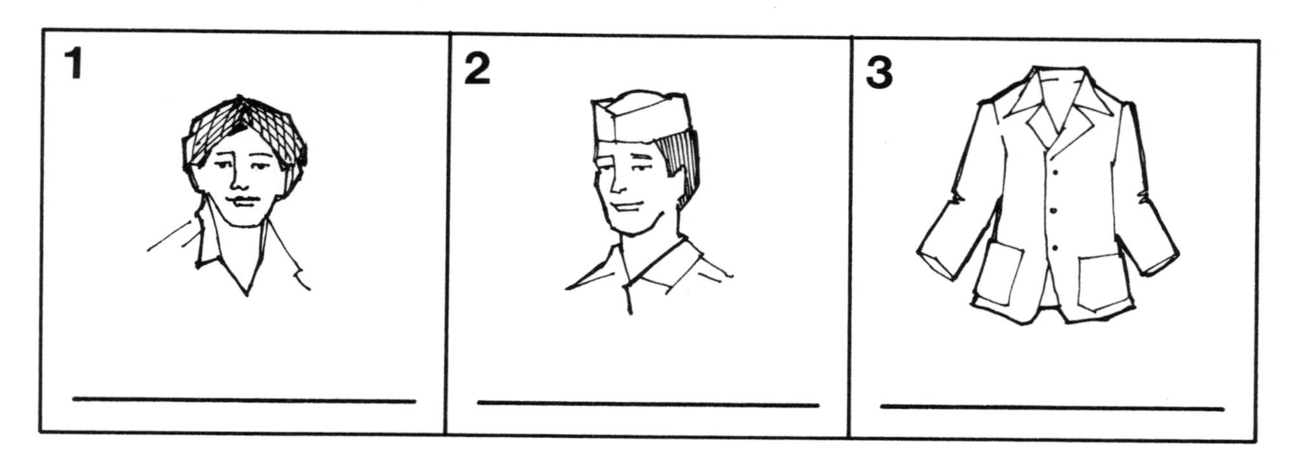

READ THE SAFETY SIGN AND POSTERS.
Answer the questions.

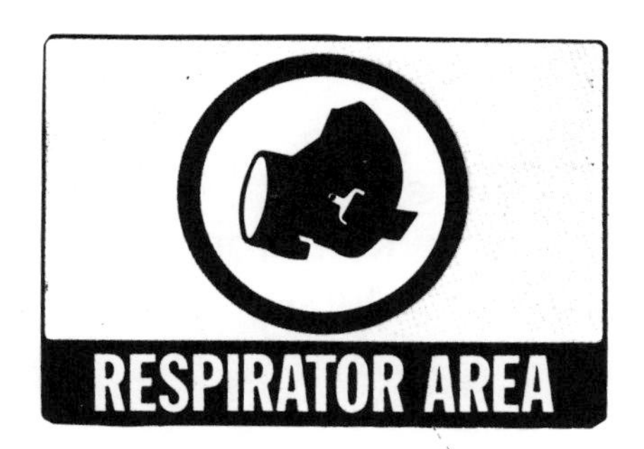

1. What should you wear?

   ____

   Why?

   ____

2. What should you wear?

   ____

   Why?

   ____

3. What should you wear?

   ____

   Why?

   ____

READ ABOUT SAFETY.

There are many different kinds of safety clothing. Different jobs require different kinds of safety clothing. Here are some examples of gloves.

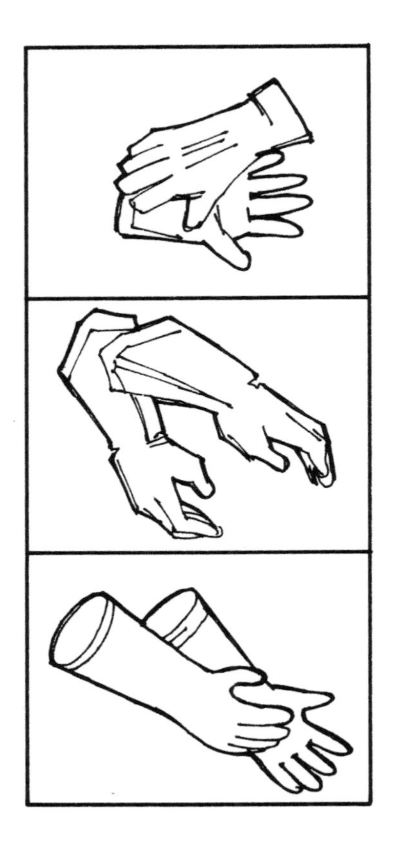

These are leather gloves.
These gloves protect workers from wood slivers and sharp metal edges.

These are gauntlet gloves.
These gloves protect welders from heat and sparks.

These are rubber gloves.
These gloves protect chemical workers from chemical burns.

CHECK YOUR UNDERSTANDING.

1. Why do workers wear leather gloves?

____

2. Why do welders wear gauntlet gloves?

____

3. Why do chemical workers wear rubber gloves?

____

4. Think of a job. What kind of gloves do the workers wear?

____

READ ABOUT SAFETY.

Here are some examples of different kinds of foot protection.

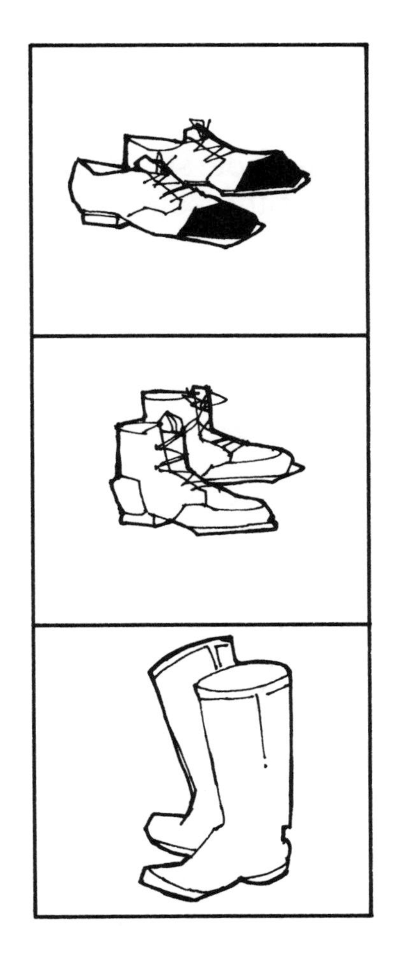

These shoes have steel toes.
Many workers wear them.
These shoes protect workers' toes.

These shoes have high tops.
Welders often wear these shoes.
These shoes protect workers from burns.

These are rubber boots.
Chemical workers often wear them.
These boots protect workers' feet from chemicals and water.

CHECK YOUR UNDERSTANDING.

1. Why do workers wear shoes with steel toes?

____

2. Why do some workers wear rubber boots?

____

3. Why do welders wear shoes with high tops?

____

4. Think of a job. What kind of shoes do the workers wear?

____

ROLE-PLAY.

Look at the pictures. Read your role. Close your book. Role-play.

A

You are the supervisor. You are talking to a new worker. You tell the worker what safety clothing to wear. You explain why the safety clothing is important. You explain about proper dress.

B

You are a new worker. You ask why you need safety clothing. You indicate you understand by saying the names of the safety clothing.

FOLLOW-UP.

Visit a workplace or shop class. What safety clothing do people wear?

LESSON FIVE: GETTING SAFETY CLOTHING

LISTEN.

The manager is talking to Stefan.

Manager: Hey, Stefan! Where are your safety glasses?

Stefan: I forgot them. They're at home. Where can I get some?

Manager: There's an extra pair in the office.

Stefan: OK.

Manager: I don't want you operating the machine when you aren't wearing safety glasses or goggles. A piece of metal could injure your eyes. ALWAYS WEAR YOUR SAFETY GLASSES!

Stefan: OK, I won't forget.

CHECK YOUR UNDERSTANDING.

1.	Stefan operates a machine.	RIGHT	WRONG
2.	Stefan forgot his safety shoes.	RIGHT	WRONG
3.	Stefan left his glasses in the office.	RIGHT	WRONG
4.	A piece of metal could hurt his eyes.	RIGHT	WRONG
5.	Stefan doesn't need safety glasses.	RIGHT	WRONG

PRACTICE.

Practice these conversations with your partner.
Practice conversations using the pictures below.

Worker:	Excuse me, I need some safety glasses.
Manager:	Yes, here you are.
Worker:	Thank you.
Manager:	You're welcome.

Worker:	Where can I get some safety glasses?
Co-worker:	Go to the office. Ask the manager.
Worker:	OK.

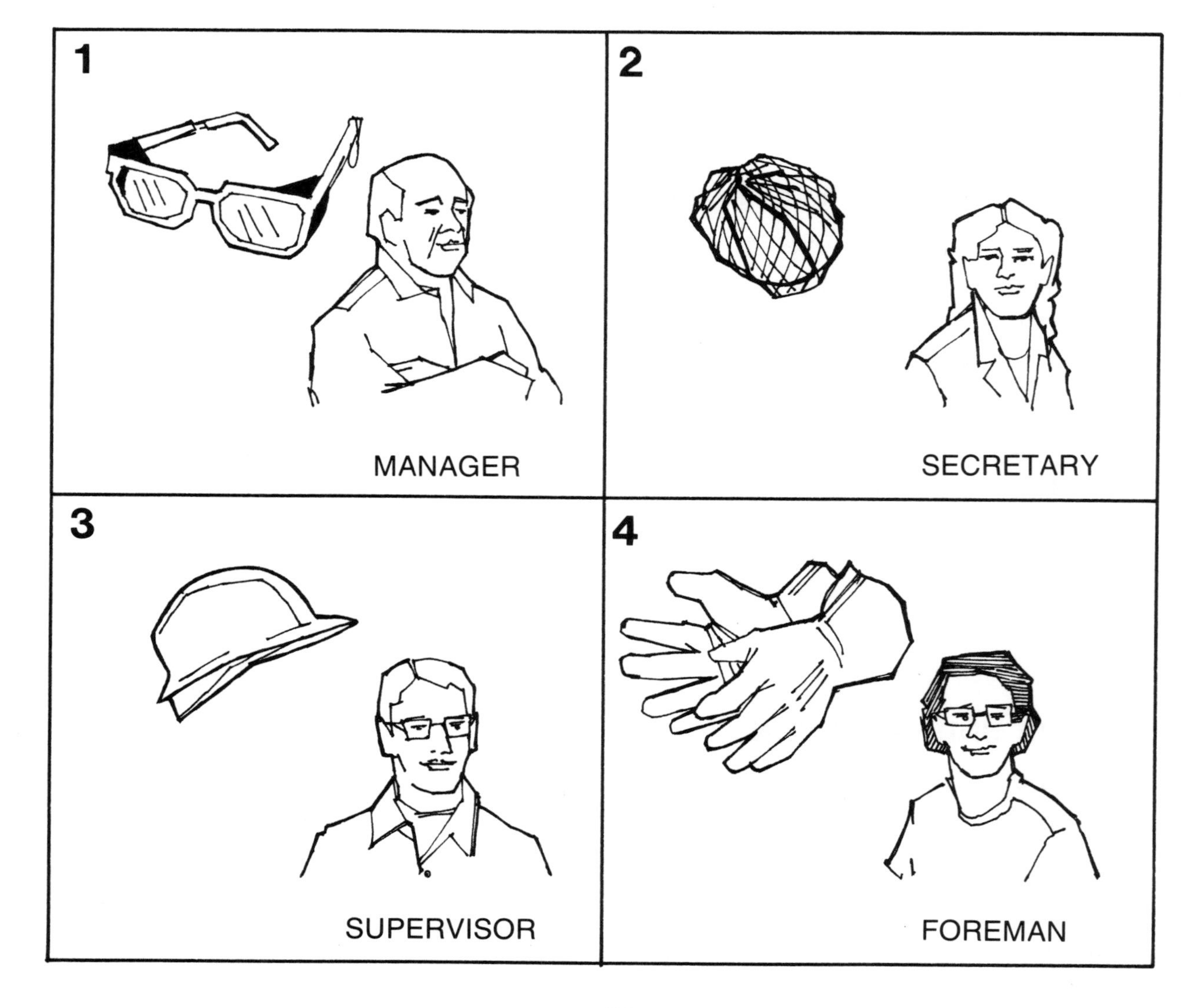

READ THE SAFETY SIGNS.

Answer the questions.

CAUTION
WEAR EYE PROTECTION
WHILE OPERATING

1. What should you wear?

____

____

Wear Goggles WHEN USING THIS TOOL

2. What should you wear?

____

____

READ ABOUT SAFETY.

Each company has its own rules about safety clothing. Some companies provide their workers with safety clothing. They give safety glasses, gloves, and other safety clothing to the workers.

At other companies, workers must buy their own safety clothing. Sometimes a company will pay a part of the cost.

It is important to find out about a company's rules. You should ask about safety clothing when you apply for a job.

CHECK YOUR UNDERSTANDING.

1. Do all companies have the same rules about safety clothing?

____

2. How do you find out about a company's rules?

____

____

READ ABOUT SAFETY.

Rosa works part-time at a cookie factory. Here is the company policy about clothing.

> COMPANY POLICY
>
> 1. The company will provide aprons. Take an apron when you arrive. Return the apron at the end of your shift so that it can be cleaned.
> 2. The company will give you one hairnet. If you need another, you must go to personnel and buy it. The cost is 25¢. REMEMBER: YOU MUST WEAR A HAIRNET AT ALL TIMES!
> 3. Each worker must wear safety shoes. You must buy your own. Show the receipt to personnel and the company will pay you 50% of the price of the first pair. If you buy more shoes, you have to pay the full cost.

CHECK YOUR UNDERSTANDING.

1. What clothing does the company give to Rosa? ______________

2. What should Rosa do if she loses her hairnet? ______________

3. Rosa buys her first pair of safety shoes. They cost $50.00.
   How much does the company pay? ______________

   How much does she pay? ______________

4. Rosa buys a second pair of safety shoes. They cost $50.00.
   How much does the company pay? ______________

   How much does Rosa pay? ______________

ROLE-PLAY.

Read your role. Close your book. Role-play with your partner.

1. Two workers are talking.

A

You left your mask at home. You want to borrow a mask from your co-worker.

B

You have an extra mask. You lend it to your co-worker.

2. A supervisor is talking to an assembler.

A

You are the supervisor. It is important that everyone wear safety glasses. One assembler does not have safety glasses. You ask the assembler why he or she isn't wearing glasses. There is an extra pair of glasses in the office.

B

You are the assembler. You have to wear safety glasses. You lost your glasses. You ask the supervisor for glasses.

3. Two workers are talking.

A

You cannot find your gloves. It is very important that you wear your gloves. You ask your co-worker for help.

B

You only have one pair of gloves. You need them. There are always extra gloves in the office.

FOLLOW-UP.

Visit a workplace or shop class. Which safety clothing is provided? What happens if a worker forgets his safety clothing?

LESSON SIX: INSISTING!

LISTEN:
Bee Vang is talking to a co-worker.

Co-worker:	Bee, carry those boxes outside for me, would you?
Bee:	OK, but I need a hard hat.
Co-worker:	Why?
Bee:	A box might fall off the forklift. I want to be safe.
Co-worker:	It'll be OK. I don't think you need a hard hat.
Bee:	No, I really need one. It's better to be safe than sorry.
Co-worker:	Well, there's probably one in the office.
Bee:	OK, I'll go get it. I'll be back in a minute.

CHECK YOUR UNDERSTANDING.

1.	Bee is a welder.	RIGHT	WRONG
2.	Bee doesn't need a hard hat.	RIGHT	WRONG
3.	Bee is worried about a falling box.	RIGHT	WRONG
4.	Bee is concerned about safety.	RIGHT	WRONG
5.	Bee is going to get a hard hat.	RIGHT	WRONG

PRACTICE.

Look at the pictures. What is unsafe? Circle the problem.

Practice the conversation with your partner.
Practice conversations using the pictures below.

Worker:	I need some ear plugs.
Supervisor:	Why?
Worker:	There's a lot of noise.
Supervisor:	It's OK.
Worker:	No, I really need ear plugs.

PRACTICE.

Look at the pictures. What could happen?

Practice the conversation with your partner.
Practice conversations using the pictures below.

Worker:	I need some gloves.
Supervisor:	Why?
Worker:	I could get cut.

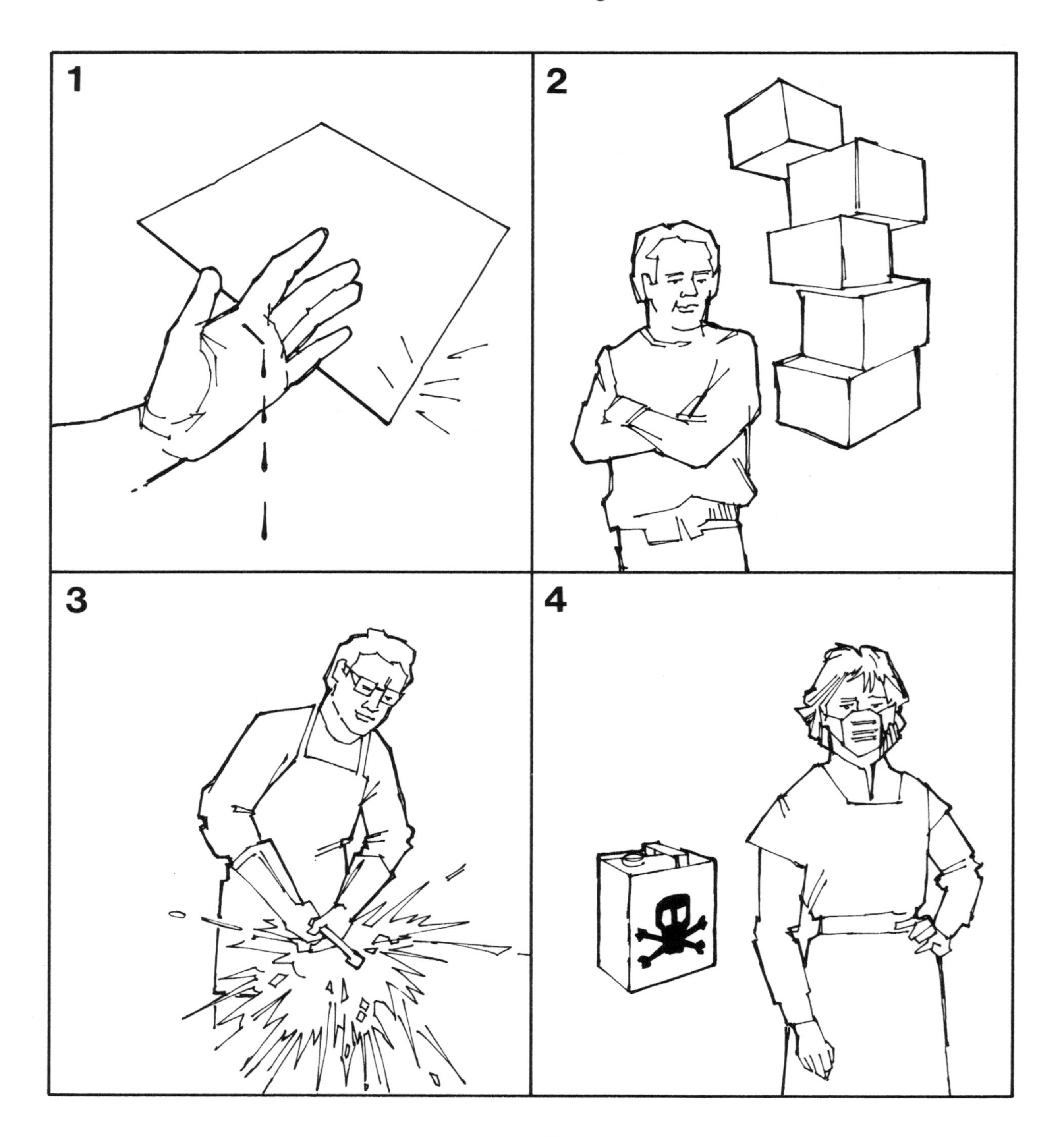

READ THE SAFETY POSTER.

Answer the question.

the safest dressed worker on the job...

wears protective equipment

What does a safe worker wear?

____

____

READ ABOUT SAFETY.

Safety is very important. Employers are interested in the safety of their workers. Employers make safety rules. Workers have to follow the rules. If workers do not follow the rules, they could lose their jobs!

Workers should also be interested in their own safety. Before a worker starts a job, he should ask himself:

1. Is the job dangerous?
2. What kind of accident and injury could happen?
3. How can I prevent an accident?
4. How can I prevent an injury?

REMEMBER: Safety clothing can protect you. If you need safety clothing, ask for it. You can ask a co-worker, supervisor, or manager.

CHECK YOUR UNDERSTANDING.

1. What could happen if a worker does not follow the safety rules?

____

2. If you need safety clothing, what should you do?

____

____

ROLE-PLAY.

Read your role. Close your book. Role-play with your partner.

1. A worker is talking to the supervisor.

A

You are the worker. There is a lot of dust. You ask the supervisor for a mask.

B

You are the supervisor. No one in the department wears a mask. You listen to the worker. You tell the worker to go to the office and get a mask.

2. A worker is talking to the boss.

A

You are the worker. There is a lot of noise. You get headaches from the noise. You ask the boss for ear plugs.

B

You are the boss. You don't think the noise is bad. You don't have any ear plugs. There might be some in the office.

3. A worker is talking to the supervisor.

A

You are the worker. You use many dangerous cleaning materials. You are concerned about burning your hands. You ask the supervisor for gloves.

B

You are the supervisor. You listen to the worker. There are extra gloves in the office.

FOLLOW-UP.

Visit a workplace or shop class. What kinds of accidents happen?
What kinds of safety clothing prevent injuries?

SAFETY PROCEDURES

LESSON SEVEN: PROPER BEHAVIOR

You will learn to understand and give instructions for proper behavior.

LESSON EIGHT: LIFT! MOVE! STACK!

You will learn to follow and give instructions for proper lifting, moving, and stacking.

LESSON NINE: DANGEROUS MATERIALS

You will learn to ask about and explain safety procedures for handling dangerous materials.

LESSON TEN: CAUTIONS

You will learn to understand and give cautions.

LESSON SEVEN: PROPER BEHAVIOR

LISTEN.

Chun Lee is watching two students. His shop teacher is talking to them.

Shop Teacher:	Hey, what's going on? No horsing around! Get back to work!
Students:	OK.
Shop Teacher:	Be careful! No horseplay in the shop! You could get hurt.

TEN MINUTES LATER

Shop Teacher:	What's going on now? Either stop playing around or get out of class. Shop can be dangerous!
Student:	Sorry, we'll get back to work.

CHECK YOUR UNDERSTANDING.

1.	Horseplay is dangerous.	RIGHT	WRONG
2.	The teacher is angry.	RIGHT	WRONG
3.	Horseplay is safe in shop class.	RIGHT	WRONG
4.	Horseplay is safe on the job.	RIGHT	WRONG

PRACTICE.

Look at the pictures. What is unsafe? Circle the problem.

Practice the conversations with your partner.

Practice conversations using the pictures below.

Foreman: Don't look out the window.
Worker: OK.

Manager: Stop looking out the window!
Worker: Sorry.

Worker A: I think you should stop looking out the window.
Worker B: Yes, I guess you're right.

READ THE SAFETY SIGN AND POSTERS.

Answer the questions.

1. What shouldn't you do?

____

____

2. What should you do?

____

Why?

____

3. What should you do?

____

____

READ ABOUT SAFETY.

Proper behavior on the shop floor or at the work-site is very important. People cause 90% of all accidents!

Here are some rules for proper behavior.

1. NO horseplay. Do NOT play around on the job.
2. NO pushing or shoving.
3. Pay attention to your work. Do NOT daydream.
4. Always watch your machine and your work.
5. Stay alert. It isn't safe to be tired or sleepy when you are working.
6. Do NOT get angry. Always "stay cool."
7. Keep your work area clean. Do NOT leave things on the floor.
8. Wipe up all spills.
9. NEVER drink alcohol on the job or before you come to work.

You can prevent accidents by following the rules.

CHECK YOUR UNDERSTANDING.

1.	People cause many accidents.	RIGHT	WRONG
2.	Horseplay is OK on the job.	RIGHT	WRONG
3.	Horseplay is OK in shop class.	RIGHT	WRONG
4.	It is OK to be tired on the job.	RIGHT	WRONG
5.	It is good to get angry.	RIGHT	WRONG
6.	Daydreaming prevents accidents.	RIGHT	WRONG
7.	Staying alert is important.	RIGHT	WRONG
8.	It is safe to leave spills on the floor.	RIGHT	WRONG
9.	It is OK to drink alcohol when you are working.	RIGHT	WRONG
10.	It is OK to drink alcohol before you come to work.	RIGHT	WRONG

ROLE-PLAY.

Look at the pictures. What is unsafe? Circle the problem.

Look at the pictures. Read your role. Close your book. Role-play with your partner.

A

You are the worker in the picture. You listen to your partner. You tell your partner that you understand.

B

You are a worker. Your partner is a worker in the picture. You tell your partner what to do or what not to do.

FOLLOW-UP.

Visit a workplace or shop class. What are the rules for proper behavior?

LESSON EIGHT: LIFT! MOVE! STACK!

LISTEN.

Marta's supervisor is talking to her.

Supervisor:	Marta, the cleaning supplies are here. Please move them to the storeroom and stack the boxes.
Marta:	OK.
Supervisor:	No, don't lift the box that way! You'll hurt your back. Bend your knees. Pick each box up slowly and carefully.
Marta:	Like this?
Supervisor:	Yes. Now that one's too heavy. Don't carry it. Get a hand truck. And be careful.
Marta:	OK.

CHECK YOUR UNDERSTANDING.

1.	Marta has to move the cleaning supplies.	RIGHT	WRONG
2.	Marta could hurt her back.	RIGHT	WRONG
3.	Marta needs a forklift truck.	RIGHT	WRONG
4.	The supervisor should move the boxes.	RIGHT	WRONG

PRACTICE.

Follow the instructions your teacher gives you.

Look at the pictures.
Give instructions to lift, move, and stack to your partner.

PRACTICE.

Look at the pictures. Practice the names.

Practice the conversations with your partner.
Practice conversations using the pictures below.

Supervisor: That's heavy. Get a hand-truck.
Worker: A hand-truck. OK.

Worker A: Be careful. Maybe you should use a hand-truck.
Worker B: OK.

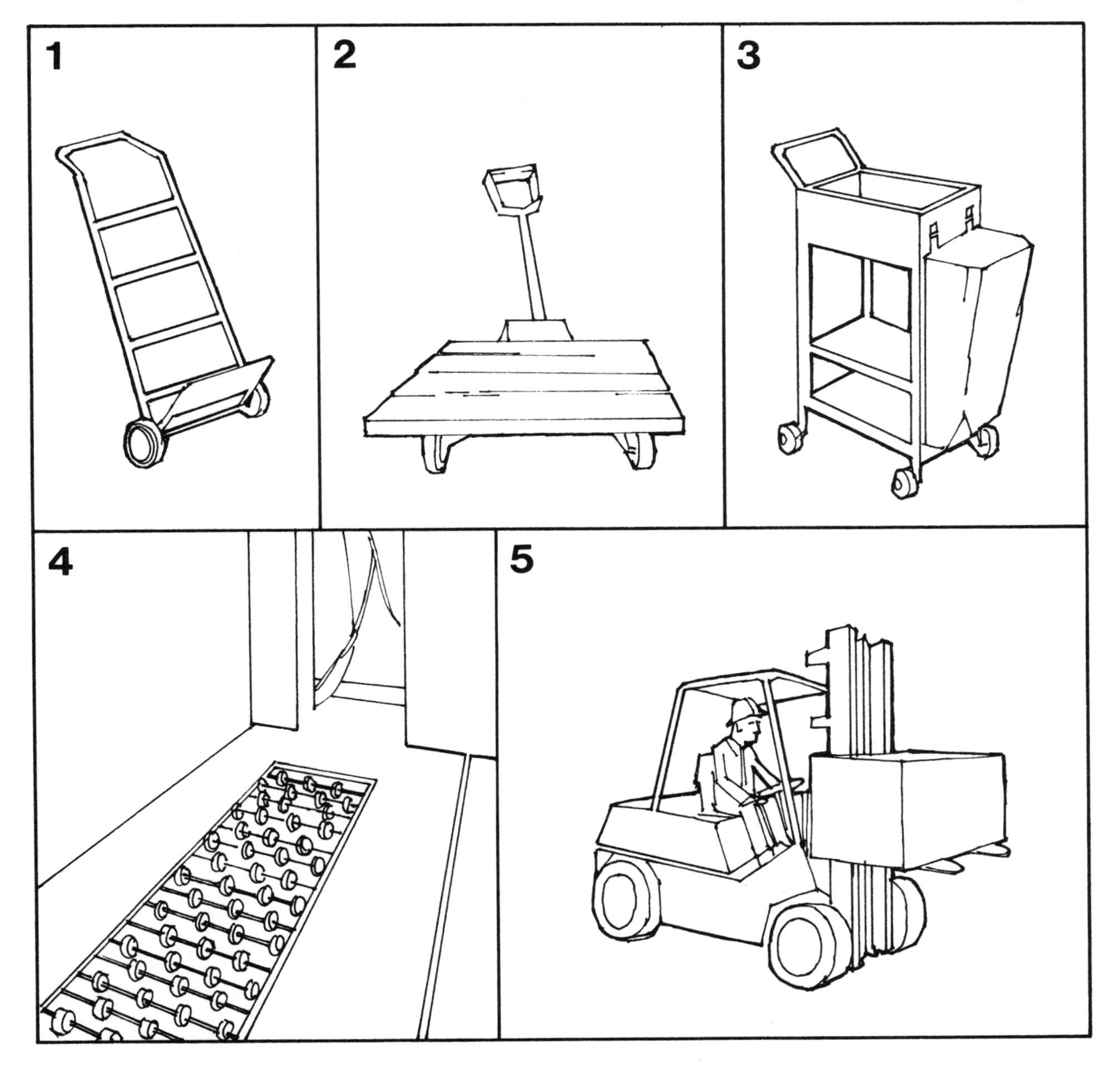

READ THE SAFETY POSTERS.

Circle the worker who is working safely.
Put an X on the worker who is not working safely.

READ ABOUT SAFETY.

Each year more than 500,000 workers are injured because they do not lift, move, or stack materials properly.

Here are some rules for proper lifting, moving, and stacking.

LIFTING:

1. Do NOT lift a heavy object by yourself.
   Get someone to help or use a mechanical aid.
2. Lift with your legs – NOT with your back.

MOVING:

1. It is better to push a load than to pull it.

STACKING:

1. Use a good, strong ladder.
2. Set the ladder properly.
3. Climb slowly and carefully.
4. Cross-tie when possible.

CHECK YOUR UNDERSTANDING.

1. Always lift things by yourself.	RIGHT	WRONG
2. It is safer to push than to pull a load.	RIGHT	WRONG
3. Climb a ladder carefully.	RIGHT	WRONG

ROLE-PLAY.

Look at the pictures. What is unsafe? Circle the problem.

Read your role. Close your book. Role-play with your partner.

A

You are a worker. Your partner is the worker in the picture. You tell you partner what he or she is doing wrong, and explain the safe way to lift, move or stack.

B

You are the worker in the picture. You listen to your partner's instruction. Thank your partner.

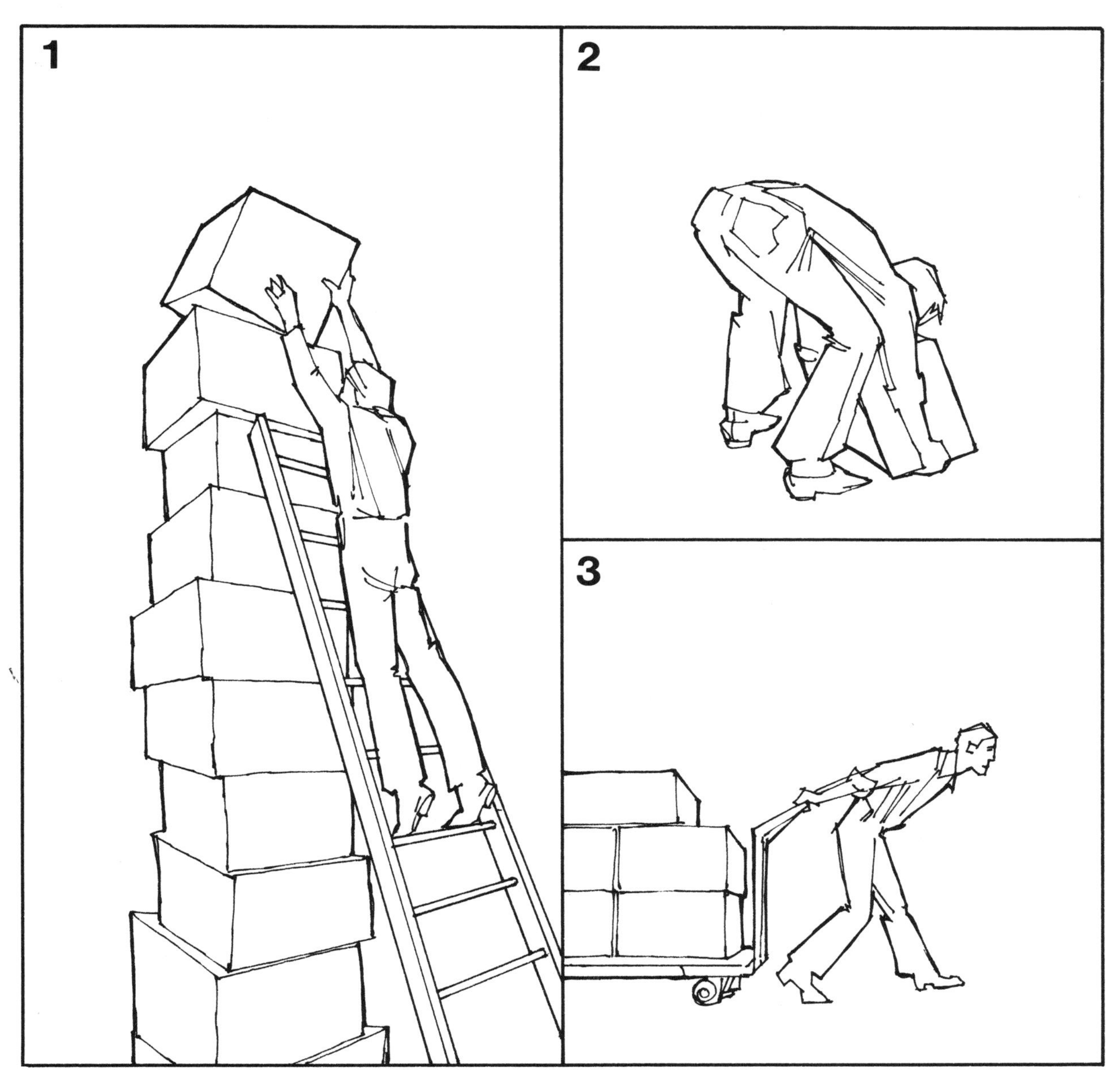

FOLLOW-UP.

Visit a workplace or shop class. What do people lift, move and stack? What mechanical aids are available?

LESSON NINE: DANGEROUS MATERIALS

LISTEN.

Johanis is talking to a co-worker.

Johanis:	Excuse me. What's this?
Co-worker:	It's paint remover. Be careful with it.
Johanis:	Why? What could happen?
Co-worker:	The fumes could make you sick.
Johanis:	How should I use it?
Co-worker:	Keep it closed when you aren't using it. Try not to breathe the fumes. Use it only when the room is well-ventilated. Oh yes, don't get it on your skin or clothes.
Johanis:	OK, I'll be careful.

CHECK YOUR UNDERSTANDING.

1.	They are talking about paint.	RIGHT	WRONG
2.	They are talking about paint remover.	RIGHT	WRONG
3.	Paint remover isn't harmful.	RIGHT	WRONG
4.	Paint remover can make you sick.	RIGHT	WRONG

PRACTICE.

Look at the pictures. What is dangerous? Put an X on the problem.

Practice the conversation with your partner.
Practice conversations using the pictures below.

Foreman:	Be careful.
Worker:	Why?
Foreman:	It's dangerous. Don't drink it.
Worker:	OK, I won't.

READ AND TALK ABOUT SAFETY.

There are many dangerous materials. Some are flammable. Some are combustible. Some are poisonous.

Flammable.

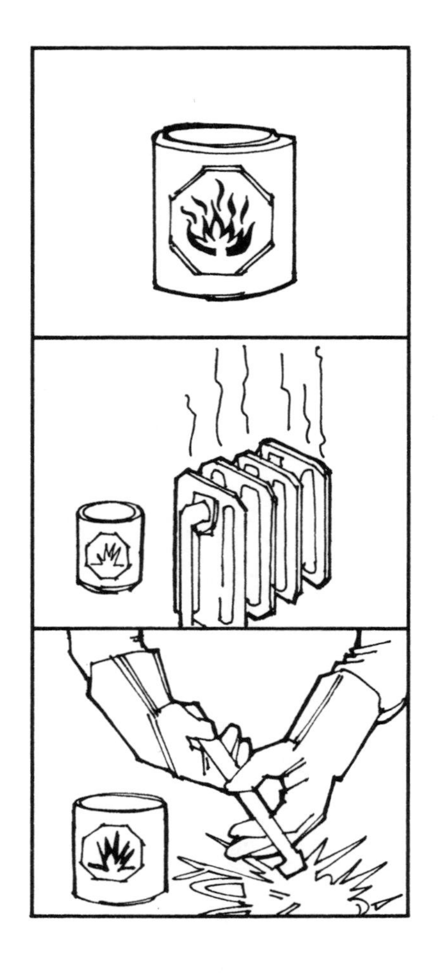

This is flammable.

It can catch on fire easily.

Do not put flammable materials near heat.

Do not put flammable materials near fire or sparks.

Combustible.

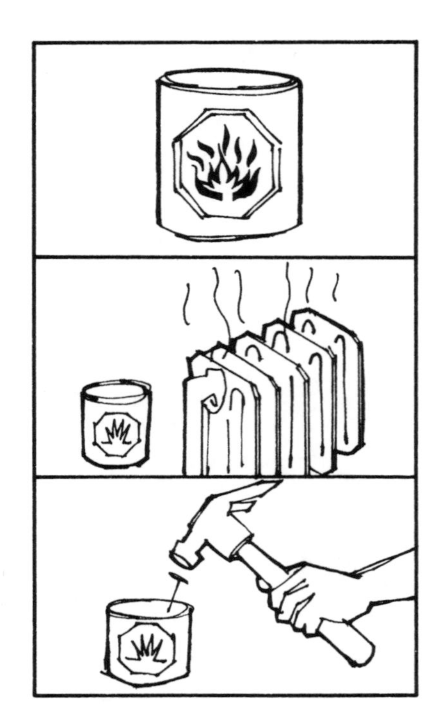

This is combustible.
It can explode.

Do not put combustible materials near heat, fire, or sparks.

Do not puncture combustible materials.

This is poisonous.
It can make you sick.
It may kill you.

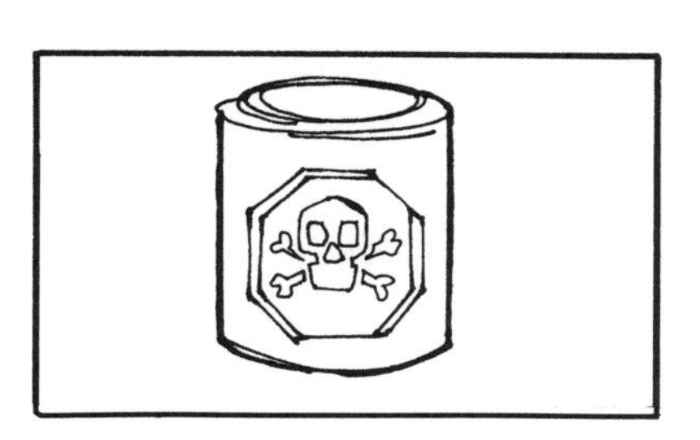

Do not eat or drink it.

Fatal if swallowed.
Do not take internally.

Do not touch it.

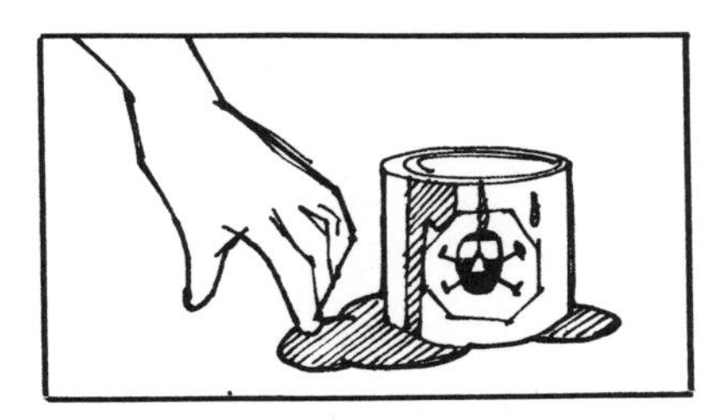

Avoid contact with skin.

Do not breathe it.

Vapor harmful.

CHECK YOUR UNDERSTANDING.

1. FATAL IF SWALLOWED means ______________________________

____

2. AVOID CONTACT WITH SKIN means ______________________________

____

3. VAPOR HARMFUL means ______________________________

____

4. DO NOT TAKE INTERNALLY means ______________________________

____

5. HARMFUL IF SWALLOWED means ______________________________

____

6 AVOID BREATHING VAPOR means ______________________________

____

7. AVOID BREATHING VAPOR AND SPRAY MIST means ______________________________

____

8. AVOID PROLONGED CONTACT WITH SKIN means ______________________________

____

READ THE SAFETY POSTERS AND SIGNS.

Answer the questions.

1. What should you do?

____

____

2. What should you do?

____

____

3. What should you do?

____

____

READ ABOUT SAFETY.

Here are some important rules about dangerous materials.

> REMEMBER!
>
> 1. Read all labels.
> 2. Wear proper clothing.
> 3. Store all dangerous materials properly.
> 4. Do NOT smoke near flammable or combustible materials.

CHECK YOUR UNDERSTANDING.

1. It is not necessary to read labels.	RIGHT	WRONG
2. It is safe to smoke near combustible materials.	RIGHT	WRONG
3. It is safe to put flammable materials near heat.	RIGHT	WRONG

CHECK YOUR UNDERSTANDING.

Read the labels. Answer the questions.

> **DANGER: HARMFUL OR FATAL IF SWALLOWED.**

1. Is this material flammable, combustible, or poisonous?

____

____

What shouldn't you do?

____

____

____

CAUTION

USE ONLY WITH ADEQUATE VENTILATION

Avoid breathing vapor and spray mist.

2. Is this material flammable, combustible, or poisonous?

____

____

What shouldn't you do?

____

____

____

W A R N I N G !!!

Contains combustible solvent.
Keep away from heat, fire, sparks, etc.

3. Is this material flammable, combustible, or poisonous?

____

____

What shouldn't you do?

____

____

____

DANGER

Contents are COMBUSTIBLE. Keep away from heat and open flame.
VAPOR HARMFUL. Use only with adequate ventilation. Avoid prolonged contact with skin. Wash hands after using.
Harmful if swallowed.

KEEP OUT OF REACH OF CHILDREN.

4. Is this material flammable, combustible, or poisonous?

____

____

What shouldn't you do?

____

____

____

ROLE-PLAY.

Read your role. Close your book. Role-play with your partner.

A

You are a worker. You are looking at a dangerous material. You ask your partner how you should use it. You repeat the instructions and thank your partner for the explanation.

B

You are a worker. You read the label on the container and answer your partner's questions.

DANGER.
FLAMMABLE MIXTURE.
DO NOT USE NEAR FIRE OR FLAME.

DANGER! Extremely flammable.
Keep away from heat, sparks, and open flame. Vapor harmful.

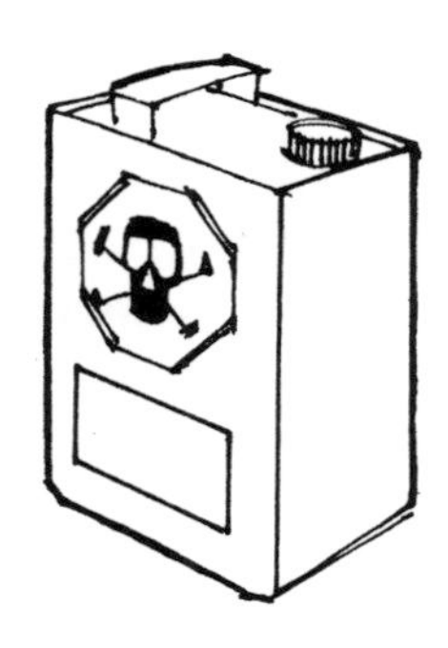

CAUTION!
Do not take internally.
Close container after each use.
KEEP OUT OF REACH OF CHILDREN

FOLLOW-UP.

Visit a workplace or shop class. What dangerous materials are there?

LESSON TEN: CAUTIONS

LISTEN.

Luis is smoking.

Worker:	Hey, Luis! There's no smoking here!
Luis:	Sorry, I didn't know.
Worker:	Well, look at the sign over there! If you want to smoke, go to the cafeteria. There are too many flammable materials to smoke here.
Luis:	OK.

CHECK YOUR UNDERSTANDING.

1.	Luis is smoking in a safe place.	RIGHT	WRONG
2.	There are many flammable materials.	RIGHT	WRONG
3.	It is unsafe to smoke in the cafeteria.	RIGHT	WRONG

PRACTICE.

Look at the pictures. What is unsafe? Circle the problem.

Practice the conversations with your partner.
Practice conversations using the pictures below.

Manager:	Don't smoke here.
Worker:	OK, sorry.
Supervisor:	You can't smoke here!
Worker:	Sorry, it won't happen again.
Worker A:	You'd better not smoke here.
Worker B:	Yes, I guess you're right.

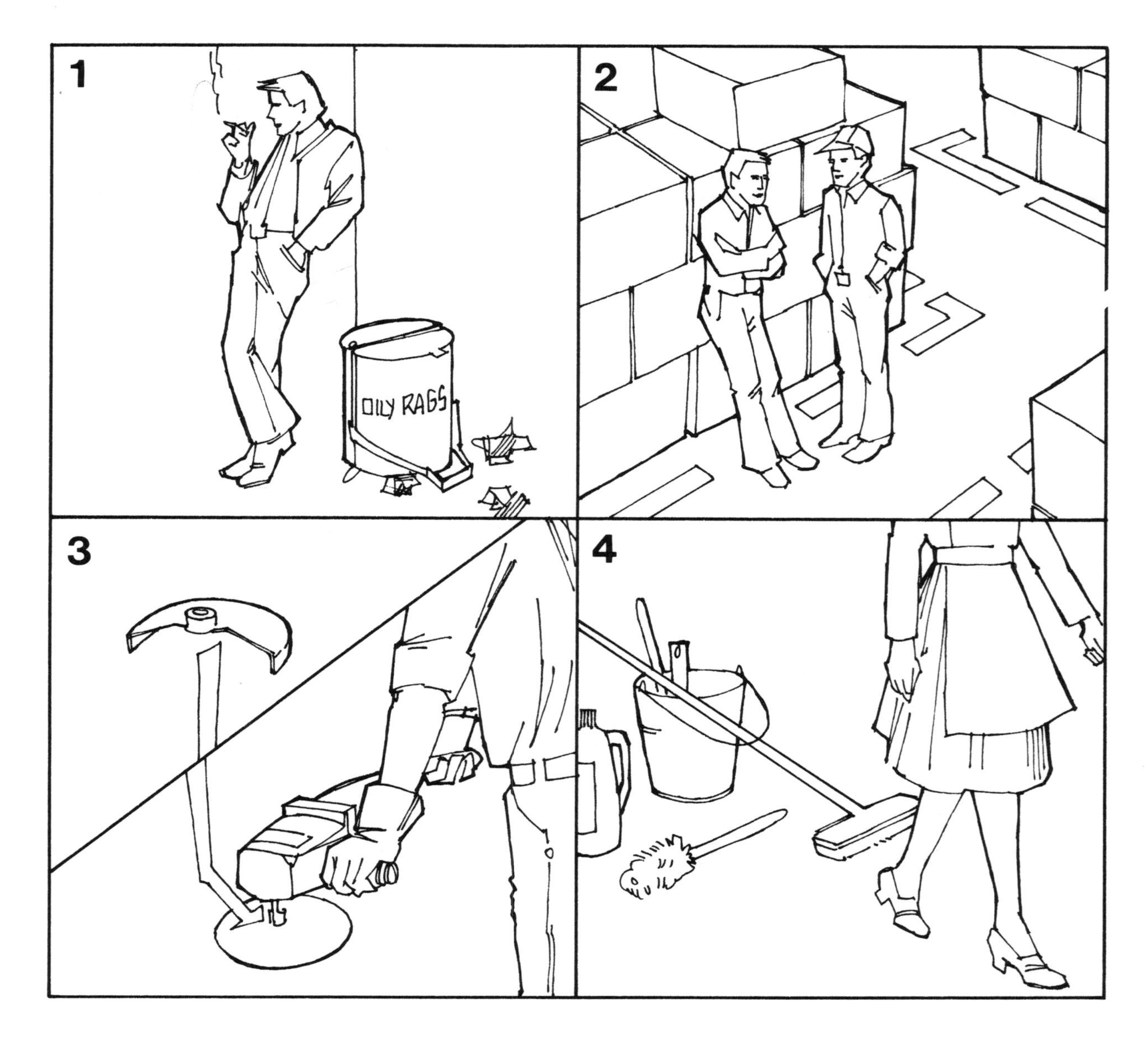

READ THE SAFETY POSTERS AND SIGNS.
Answer the questions.

1. What shouldn't you do?

____

____

Why?

____

____

2. What should you do?

____

____

Why?

____

____

3. What should you do?

____

____

Why?

____

____

READ ABOUT SAFETY.

Many accidents involve tools and machines. People must be very careful when they work with tools and machines. Here are some rules about hand tools, power tools, and machines.

HAND TOOLS

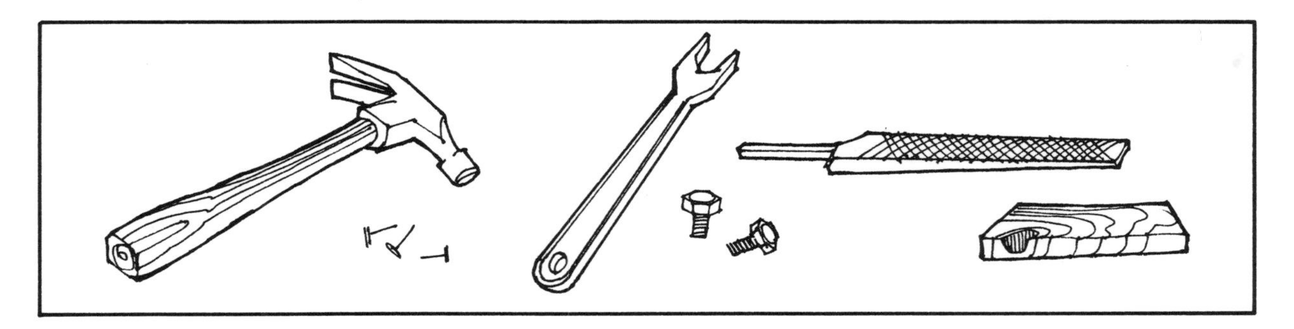

1. Use the correct tool for the job.
2. Use tools in good condition.
3. Use files with handles.
4. Do NOT carry tools in your pockets.
5. Do NOT leave tools on the floor.
6. Return tools to their proper place when you are finished.

POWER TOOLS

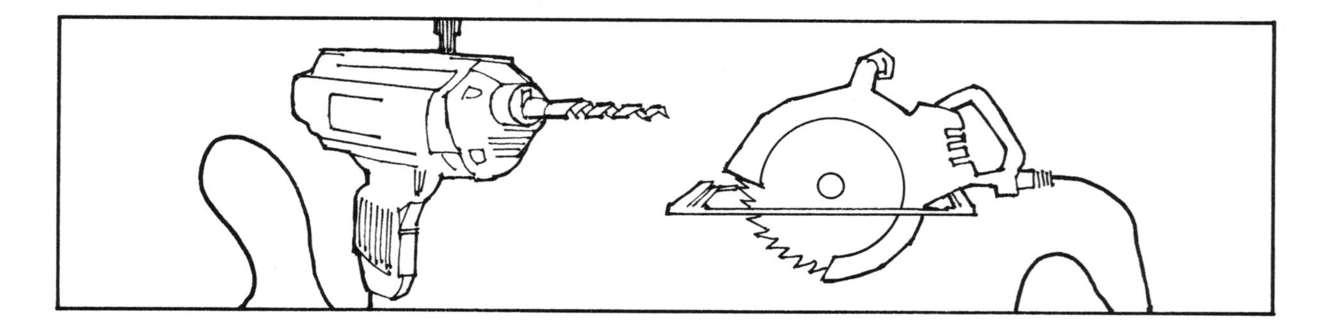

1. Never use a dull cutting edge.
2. Do NOT use a tool which has missing parts.
3. The on/off switch should work properly.
4. Machine guards should be in good condition.
5. Do NOT use a tool with a broken plug or electrical cord.
6. Tools should be grounded properly.

MACHINES

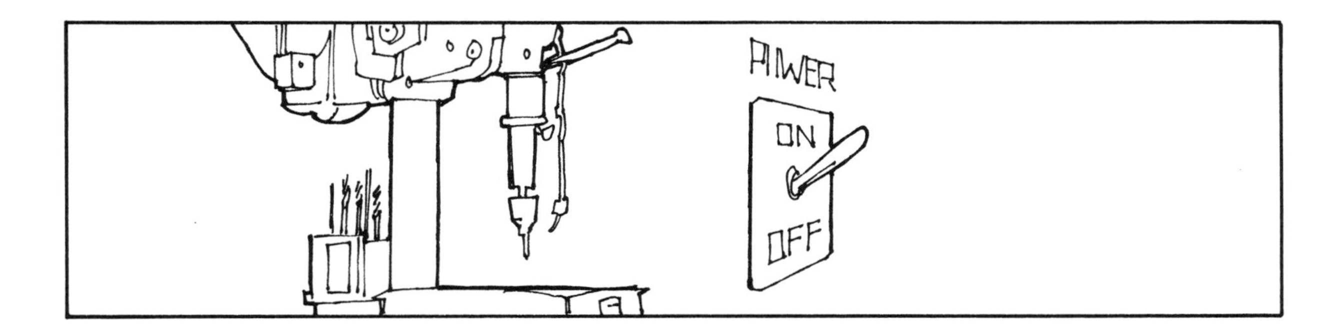

1. Check the on/off switch before starting.
2. Find the Emergency Switch before starting.
3. Keep the machine clean.
4. Always turn off the machine before cleaning or adjusting.
5. Stay away from moving parts.
6. Always turn off the machine when you are not using it.

CHECK YOUR UNDERSTANDING.

Read about the workers. Are their actions safe or unsafe?

1.	A worker carries tools in his pocket.	SAFE	UNSAFE
2.	A worker uses a tool with a broken plug.	SAFE	UNSAFE
3.	A worker puts the tools away after using them.	SAFE	UNSAFE
4.	A worker leaves tools on the floor.	SAFE	UNSAFE
5.	A worker uses a tool with a sharp cutting edge.	SAFE	UNSAFE
6.	A worker keeps the machine clean.	SAFE	UNSAFE
7.	A worker turns off the machine before cleaning it.	SAFE	UNSAFE
8.	A worker uses a file without a handle.	SAFE	UNSAFE
9.	A worker uses a machine with a broken guard.	SAFE	UNSAFE
10.	A worker uses a machine with a broken Emergency Switch.	SAFE	UNSAFE

ROLE-PLAY.

Look at the pictures. What is wrong? Circle the problem.
Read your role. Close your book. Role-play with your partner.

A

You are a worker. Your partner is the worker in the picture.
Tell your partner what to do or what not to do.

B

You are the worker in the picture. Listen to your partner. Respond appropriately.

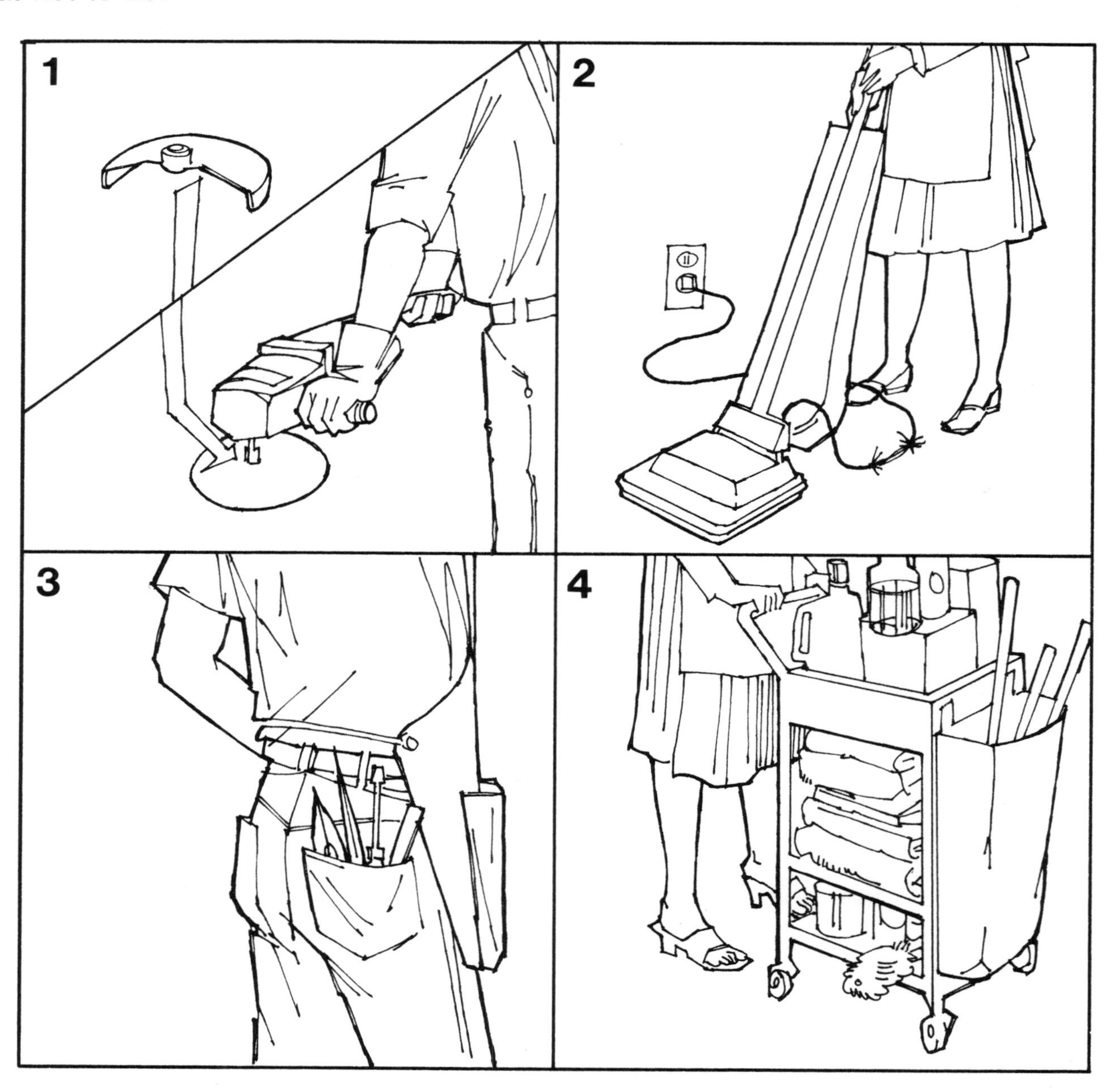

FOLLOW-UP.

Visit a workplace or shop class. What cautions do workers give?

WORKING CONDITIONS

LESSON ELEVEN: IS IT SAFE?

LISTEN.

The supervisor wants Johanis to operate a grinding machine today.

Supervisor:	Hey, Johanis. I need you to operate a grinding machine in Department 10.
Johanis:	Right now?
Supervisor:	Yeah. Come here, I'll show you. Do you know how to operate a grinder?
Johanis:	Yeah, no problem. But it's pretty dark here. There isn't a lot of light. Is it safe?
Supervisor:	We've never had an accident.
Johanis:	Well, I'm still not sure it's safe.
Supervisor:	You could be right. I'll talk to the manager. We'll see what we can do.
Johanis:	Thanks. I appreciate it. I'd rather be safe than sorry.
Supervisor:	Yeah, I agree with you.

CHECK YOUR UNDERSTANDING.

1.	Johanis is going to work in Department 10.	RIGHT	WRONG
2.	The lighting is good.	RIGHT	WRONG
3.	Johanis is worried about the lighting.	RIGHT	WRONG
4.	There have been many accidents.	RIGHT	WRONG

PRACTICE.

Look at the pictures. What's unsafe?

Practice the conversation with your partner.
Practice conversations using the pictures below.

Worker:	There's a lot of dust. Is it safe to work here?
Co-worker:	I'm not sure. Let's ask the supervisor.
Worker:	OK.

READ THE SAFETY POSTER.

Answer the question.

What should you do?

____

____

READ ABOUT SAFETY.

Accidents are sometimes caused by unsafe conditions. Every worker should look for and report unsafe conditions. Some unsafe conditions are:

1. Poor lighting.
2. Poor ventilation.
3. No safety guards on machines.
4. Unsafe electrical systems.
5. Fire and explosion hazards.
6. Crowded work-sites.
7. Machines that do not work properly.

If you are not sure about the working conditions, ask someone if it is safe!
If you think that the working conditions are not safe, report them!

CHECK YOUR UNDERSTANDING.

Think about a job. Describe safe working conditions for the job.

____

____

____

____

ROLE-PLAY.

Read your role. Close your book. Role-play with your partner.

A

You are a worker. You describe the unsafe condition and ask your partner if it is safe to work.

B

You are a worker. You tell your partner that you aren't sure if it is safe. You suggest that your partner talk to the supervisor.

A

You are a worker. You describe a possible unsafe condition to your partner.

B

You are a worker. There has never been an accident. You ask your partner what you should do.

FOLLOW-UP.

Visit a workplace or shop class. Are there any unsafe conditions? What should workers be concerned about?

LESSON TWELVE: REPORTING AN UNSAFE CONDITION

LISTEN.

Luis wants to report an unsafe condition.

Luis:	Mr. Jackson, I'd like to report an unsafe conditon.
Mr. Jackson:	Yes?
Luis:	There are a lot of chemical fumes. I'm feeling dizzy. I think the ventilation is blocked.
Mr. Jackson:	Are other people feeling sick, too?
Luis:	Yes, John had to go home earlier.
Mr. Jackson:	OK, I'll check into it. We'll see what we can do. In the meantime, make sure everyone uses a mask. We don't want people risking their health. I'll get back to you. Now maybe you should see the nurse.
Luis:	OK. Thanks.

CHECK YOUR UNDERSTANDING.

1. Luis wants to report a problem.	RIGHT	WRONG
2. Luis isn't feeling well.	RIGHT	WRONG
3. The ventilation is good.	RIGHT	WRONG
4. Everyone has to wear a mask.	RIGHT	WRONG

PRACTICE.

Look at the pictures. What is unsafe? Circle the problem.

Practice the conversation with your partner.

Practice conversations using the pictures below.

Worker:	Excuse me, I'd like to report an unsafe condition.
Supervisor:	Yes?
Worker:	The lighting is poor. Somebody might get hurt.
Supervisor:	OK, I'll look into it.

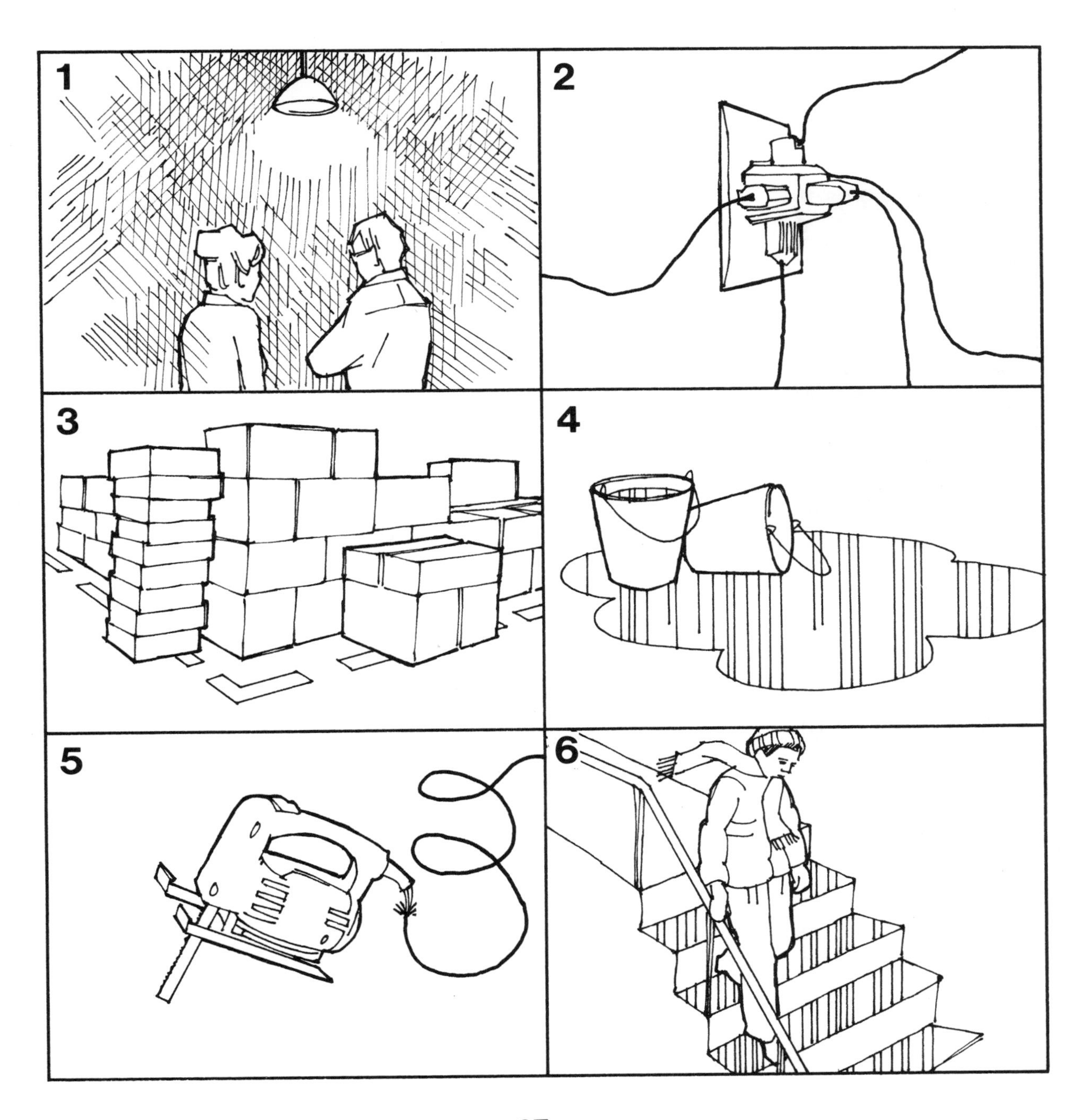

READ THE SAFETY POSTERS.

Answer the questions.

1. What is unsafe?

   ____

   What should you do?

   ____

   ____

2. What should you do?

   ____

   ____

3. What is unsafe?

   ____

   ____

   What should you do?

   ____

   ____

READ ABOUT SAFETY.

Every worker has a responsibility for the safety of all workers. If a worker finds an unsafe working condition, he or she should report it.

Each workplace has different procedures for reporting unsafe conditions. In some companies, workers report unsafe conditions to the supervisor. Some companies have special safety inspectors. They are responsible for safety in the workplace. A worker can report an unsafe condition to them. If there is a union, a worker can report an unsafe condition to the shop steward.

Many companies have special forms for reporting unsafe conditions. A worker fills out the form and gives it to the supervisor, manager, or safety supervisor.

The government also has safety inspectors. They are called O.S.H.A. inspectors. If an unsafe conditon is not made better, a worker can report the condition to them.

CHECK YOUR UNDERSTANDING.

1. Who is responsible for safe working conditions?

   ____

   ____

2. Who can workers report unsafe conditions to?

   ____

   ____

3. Do all companies have the same reporting procedures?

   ____

FILL OUT THE FORM.

Look for unsafe working conditions in the picture.
Fill out the form on page 71. Decide if each condition is safe or unsafe.
Comment about what is unsafe.

FILL OUT THE FORM.

Look at the picture on page 70. Fill out the form below. Describe the unsafe conditions.

CHECKLIST FOR WORKING CONDITIONS

CONDITION	SAFE	UNSAFE	COMMENTS
Aisles			
Electrical connections			
Floors			
Lighting			
Machines			
Safety clothing			
Storage of dangerous materials			
Storage of tools			
Worker appearance			
Worker behavior			
Other:			

Signature: ______________________ Date: ____________

ROLE-PLAY.

Look at the picture on page 70. Read your role. Role-play with your partner.

A

You are a worker. You want to report an unsafe condition to the supervisor.

B

You are the supervisor. You listen to the worker and you thank the worker for reporting the unsafe condition. You tell the worker you will check into it.

FOLLOW-UP.

Visit a workplace or shop class. What are the procedures for reporting unsafe conditions?

LESSON THIRTEEN: SUGGESTIONS

LISTEN.

Johanis is talking to the shop steward.

Johanis:	Mike, can you come here a minute?
Shop Steward:	Sure. Be right there.
Johanis:	It's pretty dark here. I don't think it's safe to operate the machines here.
Shop Steward:	Do you have a suggestion?
Johanis:	Yes, I think we need another ceiling light. It would be safer if we had more light.
Shop Steward:	OK, I'll look into it. You could be right. I'll talk to the manager today and get back to you this afternoon. Thanks for the suggestion.

CHECK YOUR UNDERSTANDING.

1.	Johanis is concerned about safety.	RIGHT	WRONG
2.	There are too many lights.	RIGHT	WRONG
3.	The working conditions could be better.	RIGHT	WRONG
4.	Johanis made a suggestion.	RIGHT	WRONG
5.	The shop steward will talk to the manager.	RIGHT	WRONG

PRACTICE.

Look at the pictures. What is unsafe? Circle the problem.

Practice the conversation with your partner.
Practice conversations using the pictures below.

Worker:	There's a lot of noise. I think it would be safer, if we had ear plugs.
Supervisor:	OK. I'll look into it.

READ THE SAFETY POSTER.

Answer the question.

What should you do?

____

____

READ ABOUT SAFETY.

It is also important to make suggestions for improving the working conditions. Management is interested in hearing your suggestions. Some companies give rewards for the best suggestions.

REMEMBER — Before you start working, you should:

1. Study the job.
2. Find out what accidents have happened or could happen.
3. Look for hazards.
4. Correct or report all hazards.

It is important to report all unsafe working conditions.

CHECK YOUR UNDERSTANDING.

1. Why would a manager be interested in your suggestions?

____

____

2. How can you prevent accidents?

____

____

FILL OUT THE FORMS.

Look at the pictures. What are the unsafe conditions?
Fill out the forms using your name and today's date.

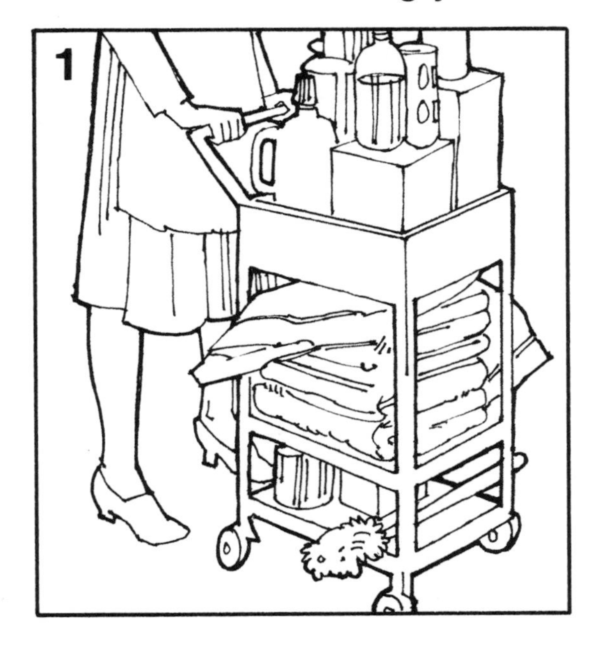

1

NAME: ____________________ DATE ________
Describe the unsafe working condition:
Suggestions:
SIGNATURE: ______________________________

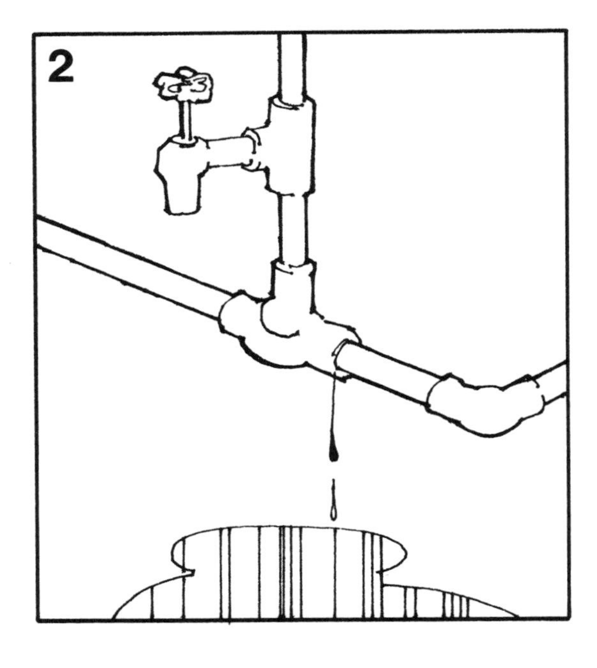

2

NAME: ____________________ DATE ________
Describe the unsafe working condition:
Suggestions:
SIGNATURE: ______________________________

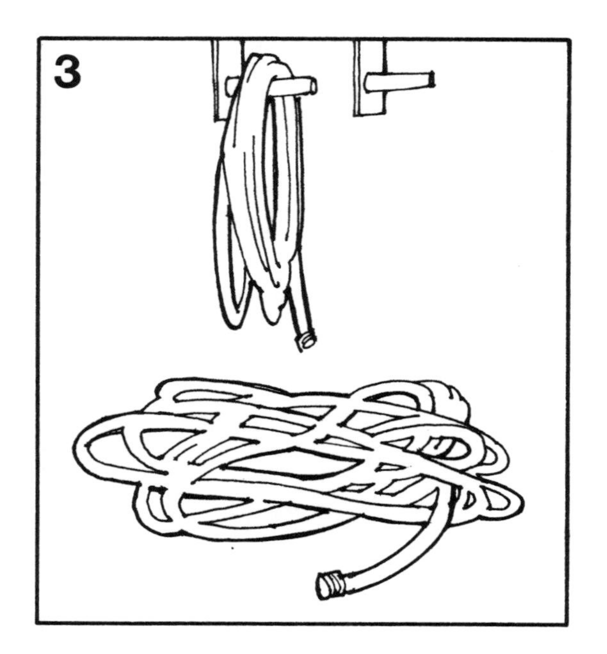

NAME: ____________________ DATE ________

Describe the unsafe working condition:

Suggestions:

SIGNATURE: ______________________________

NAME: ____________________ DATE ________

Describe the unsafe working condition:

Suggestions:

SIGNATURE: ______________________________

ROLE-PLAY.

Look at the pictures on pages 75 and 76.
Read your role. Close your book. Role-play with your partner.

1. A worker is talking to the manager.

A

You are the worker. You tell the manager that you would like to report an unsafe condition. You describe the hazard and explain what kind of accident could happen. Then you make a suggestion.

B

You are the manager. You are interested in learning about unsafe conditions. You ask for a suggestion.

2. A worker is talking to the supervisor.

A

You are the worker. You want to report an unsafe condition. You explain what kind of accident could happen and you make a suggestion.

B

You are the supervisor. You want to find out about unsafe conditions. You listen to the worker. You tell the worker that you will talk to the manager.

3. A worker is talking to the shop steward.

A

You are the worker. You want to describe an unsafe condition. You ask what the shop steward can do.

B

You are the shop steward. You listen to the worker. You ask the worker to fill out a form. You will give the form to the manager and talk to the manager about the unsafe condition. You will report back to the worker next week.

FOLLOW-UP.

Visit a workplace or shop class. What are the procedures for making suggestions?

LESSON FOURTEEN: WARNINGS

LISTEN.

Anwar sees a dangerous situation.

Anwar:	Look out! A box is falling!
Bill:	What happened?
Anwar:	A box fell. It almost hit you.
Bill:	Thanks for the warning. I could have been hurt!

CHECK YOUR UNDERSTANDING.

1. There was a dangerous situation.	RIGHT	WRONG
2. Anwar warned his friend.	RIGHT	WRONG
3. Anwar's friend got hurt.	RIGHT	WRONG

PRACTICE.

Look at the pictures. What is dangerous? Practice giving warnings.

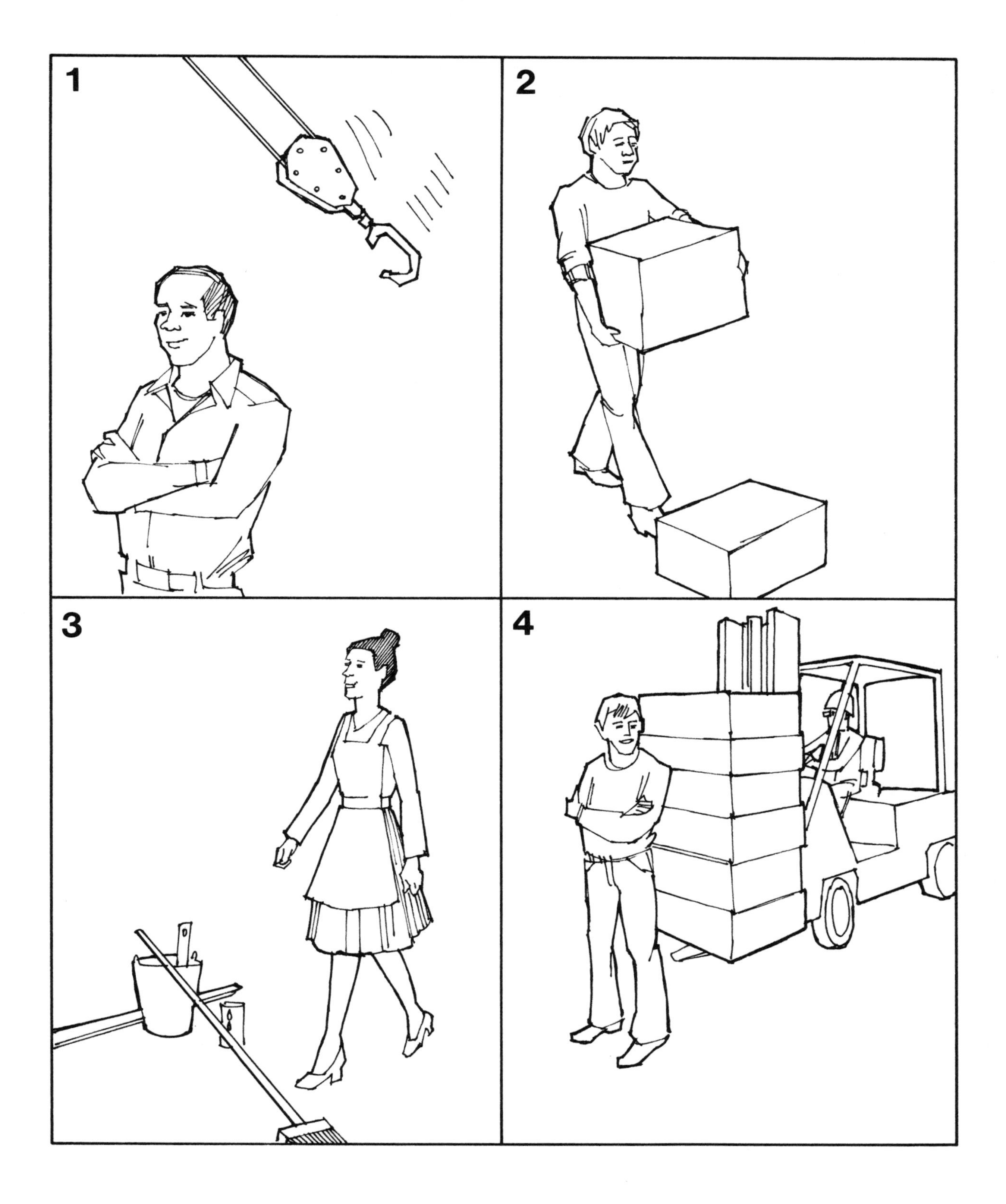

READ THE SAFETY POSTERS.

Answer the questions.

DANGER

NO SMOKING

Trouble!

1. What shouldn't you do?

____

____

DANGER

ACID

2. What is the danger?

____

____

READ ABOUT SAFETY.

Everyone should be concerned about the safety of other workers. It is important to look for dangerous situations and to warn others. Always give warnings loudly and quickly! Protect yourself and your co-workers!

CHECK YOUR UNDERSTANDING.

1. Think of a job. What is a dangerous situation? What warning would you give?

____

____

____

ROLE-PLAY.

Read your role. Close your book. Role-play with your partner.

1. Two workers are stacking boxes.

A	B
You see that a box is going to fall. Warn your co-worker.	You hear the warning and respond. You thank your co-worker for the warning.

2. Two carpenters are carrying wood outside.

A	B
You see a hole in the ground. Your co-worker might fall in the hole. Warn your co-worker.	You hear the warning and respond. You thank your co-worker for the warning.

3. Two people are working in the kitchen of a restaurant.

A	B
There is a very hot pan on the table. Your co-worker is going to pick it up. Warn your co-worker.	You hear the warning and respond. You thank your co-worker for the warning.

4. Two workers are walking in a crowded aisle.

A	B
There is a pile of sheet metal in the aisle. The edges are very sharp. Your co-worker is walking very close to the sheet metal. Warn your co-worker.	You hear the warning and respond. You thank your co-worker for the warning.

FOLLOW-UP.

Visit a workplace or shop class. What are some dangerous situations? What kinds of warnings would someone give?

UNIT FIFTEEN: FIRE!

LISTEN.

Luis sees a small fire.

Luis: FIRE! EMERGENCY!
Everybody out! Don't run!

CHECK YOUR UNDERSTANDING.

1.	Luis sees a small fire.	RIGHT	WRONG
2.	Luis asks for help.	RIGHT	WRONG
3.	Luis tells everyone to leave.	RIGHT	WRONG

PRACTICE.

Look at the pictures. What is dangerous?

Practice giving warnings for each situation.

READ THE SAFETY POSTER.

Answer the question.

What should you do?

____

____

READ THE SIGNS.

Look at the pictures. Write the correct words.

FIRE HOSE

FIRE BLANKET

FIRE EXTINGUISHER

FIRE EXIT

FIRE ALARM

FIRE DOOR – KEEP CLOSED

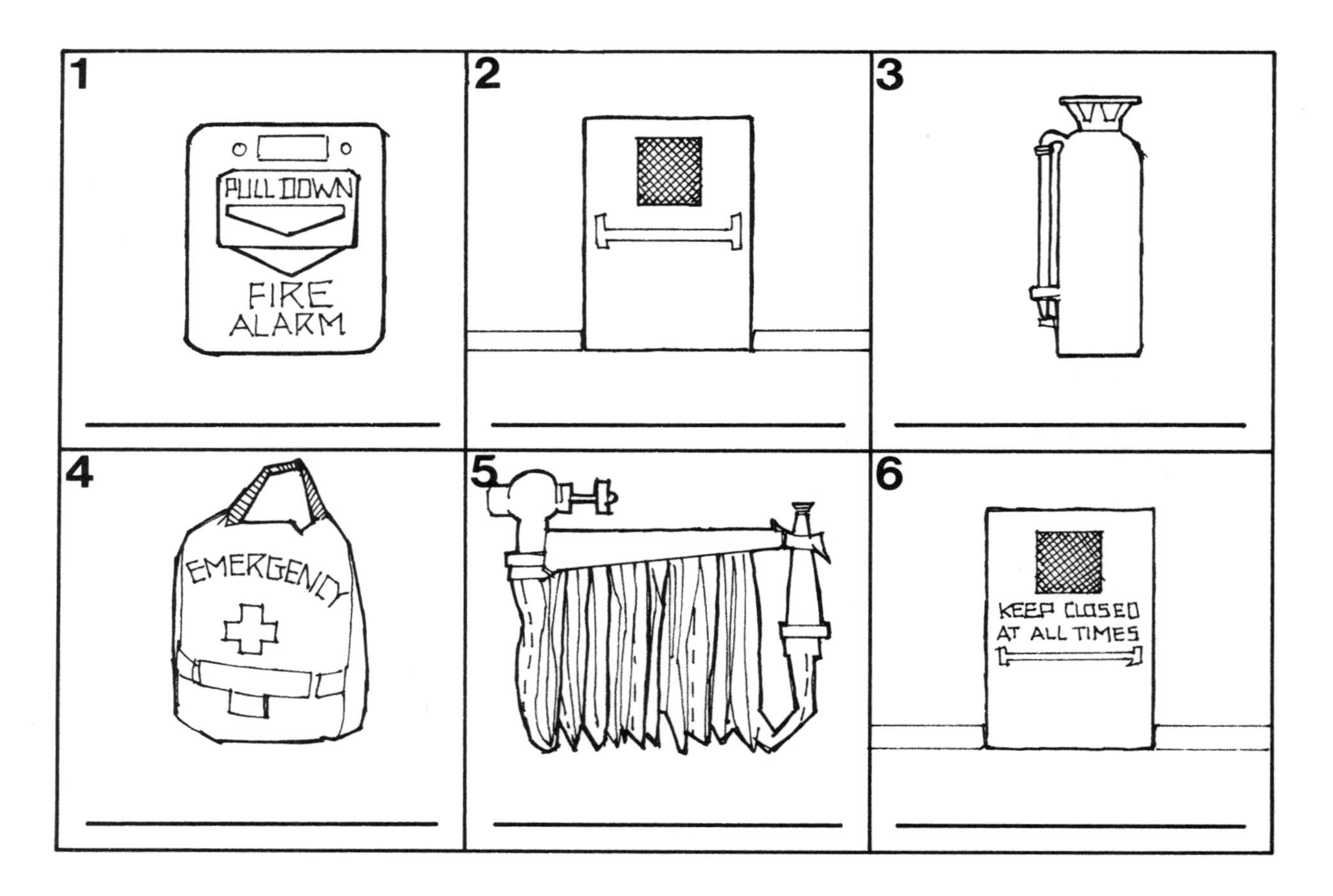

READ ABOUT SAFETY.

Preventing fires.

All workers can help prevent fires. Here are some rules.

1. Smoke only in approved places.
2. Keep the workplace free of oily or flammable rags.
3. Handle and store all flammable and combustible materials properly.
4. Check all electrical connections. Report dangerous connections immediately.
5. Keep machines in good working order.

CHECK YOUR UNDERSTANDING.

Think of a job. How can a worker prevent a fire?

____

____

READ ABOUT SAFETY.

Reporting fires.

If a fire occurs, DO NOT PANIC!

1. Report the fire — use a telephone or use a fire alarm.
2. Warn everyone. Help them get out.
3. If it is a small fire, help to put it out. DO NOT TRY TO PUT OUT A BIG FIRE!

CHECK YOUR UNDERSTANDING.

What would you do in case of a fire?

____

____

____

Using a fire extinguisher.

Fire extinguishers are used to put out small fires.

There are 3 main kinds of fire extinguishers. Each one puts out a different kind of fire. They are:

Extinguisher A	Extinguisher B	Extinguisher C
Wood	Paint	Electrical
Paper	Solvent	
Cloth	Oil	
	Other flammable liquids	

CHECK YOUR UNDERSTANDING.

Look at the fires.
Match the fires with the correct fire extinguisher.

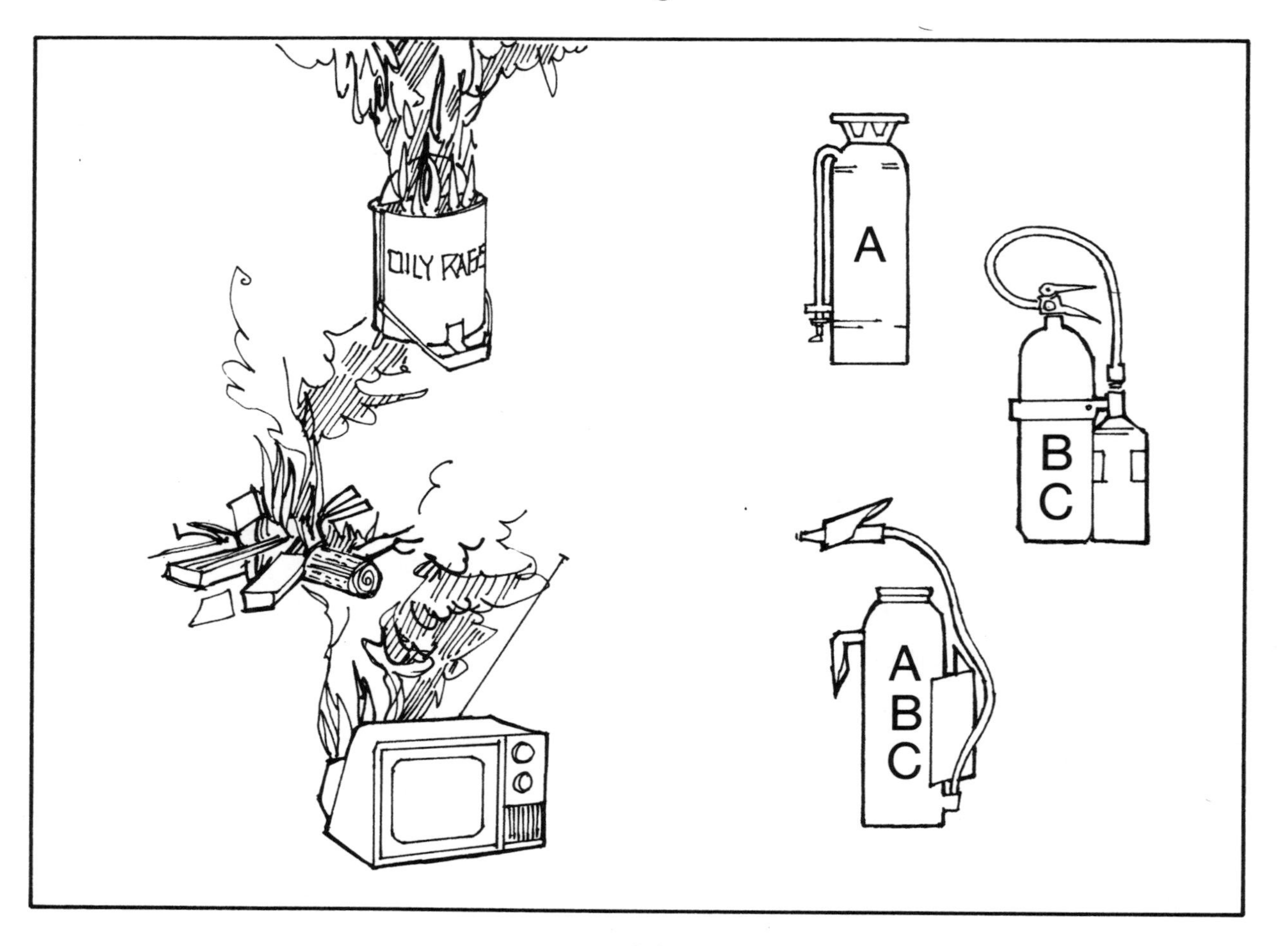

A very common fire extinguisher is the multi-purpose fire extinguisher. It can be used for all three kinds of fires.
It is easy to use.

HOLD UPRIGHT.

PULL OUT RING.

SQUEEZE HANDLE.

SWEEP FROM SIDE TO SIDE.

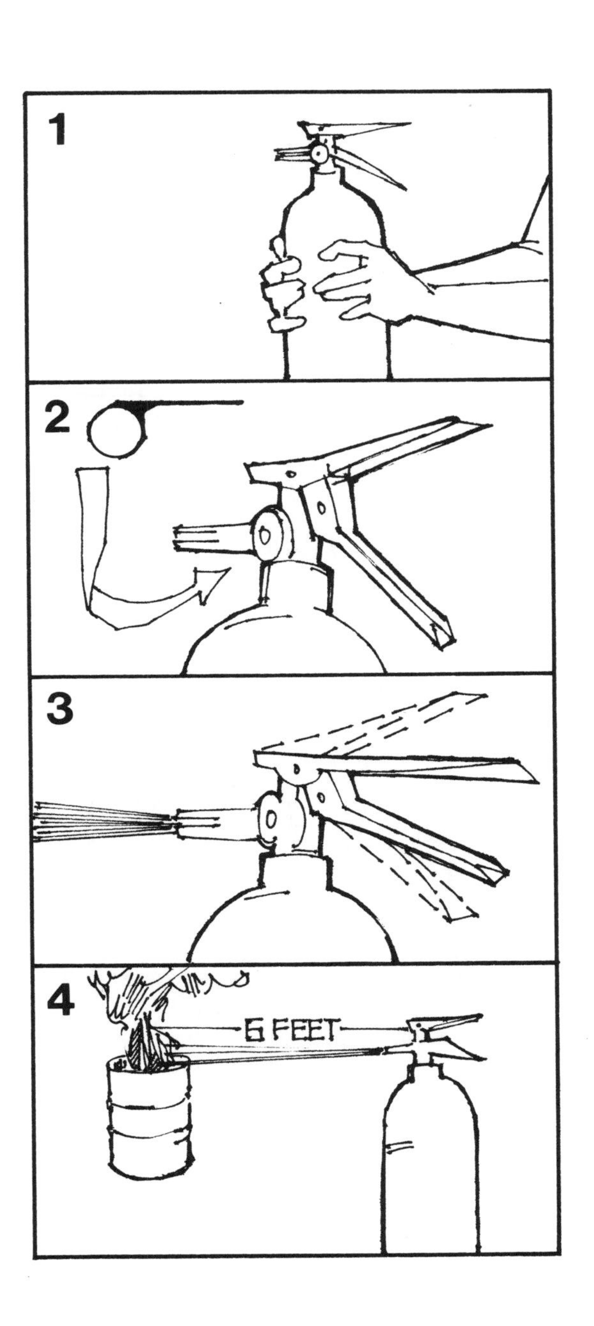

CHECK YOUR UNDERSTANDING.

Look at the pictures. Do not look at the instructions.

Explain to your partner how to operate the multi-purpose fire extinguisher.

ROLE-PLAY.

Look at the pictures. Would you put out the fire?
Practice giving warnings using the pictures below.

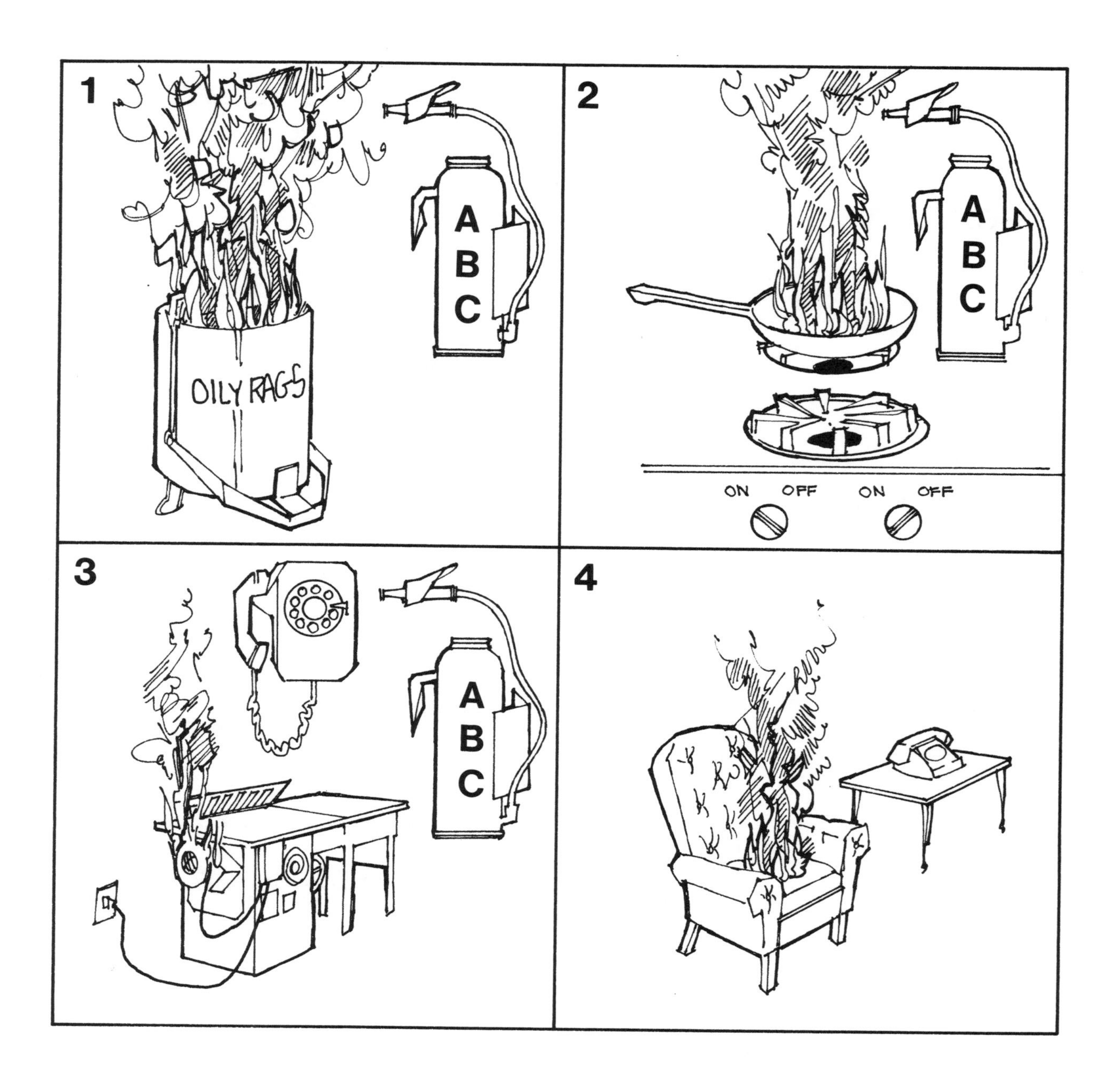

FOLLOW-UP.

Visit a shop or job-site. What kinds of fires can occur? What fire fighting equipment is there? What are the procedures for reporting a fire?

ACCIDENTS

LESSON SIXTEEN: **A MINOR ACCIDENT**

You will learn to give and accept advice politely.

LESSON SEVENTEEN: **REPORTING AN INJURY**

You will learn to report accidents and injuries.

LESSON EIGHTEEN: **HOW DID IT HAPPEN?**

You will learn to identify events which happened before an accident.

LESSON NINETEEN: **MORE SUGGESTIONS**

You will learn to suggest ways of preventing accidents.

UNIT SIXTEEN: A MINOR ACCIDENT

LISTEN.

Kiem cut herself on a piece of broken glass.

Kiem;	Oooohhhh . . . . .
Co-worker:	What happened?
Kiem:	I cut myself.
Co-worker:	You'd better go to the first aid room.
Kiem:	No, it's not too bad.
Co-worker:	It could get infected. You really should get first aid. Now go — I'll tell the supervisor.
Kiem:	Thanks.

CHECK YOUR UNDERSTANDING.

1.	Kiem broke her finger.	RIGHT	WRONG
2.	Kiem cut herself.	RIGHT	WRONG
3.	Kiem is going to the hospital.	RIGHT	WRONG
4.	Kiem is going to get first aid.	RIGHT	WRONG
5.	Kiem will report the accident to the supervisor.	RIGHT	WRONG

PRACTICE.

Look at the pictures. What happened?

Practice the conversation with your partner.
Practice conversations using the pictures below.

Co-worker:	What happened?
Worker:	I cut my hand.
Co-worker:	You should go to the first aid room. I'll report it to the supervisor.
Worker:	OK, I will. Thanks.

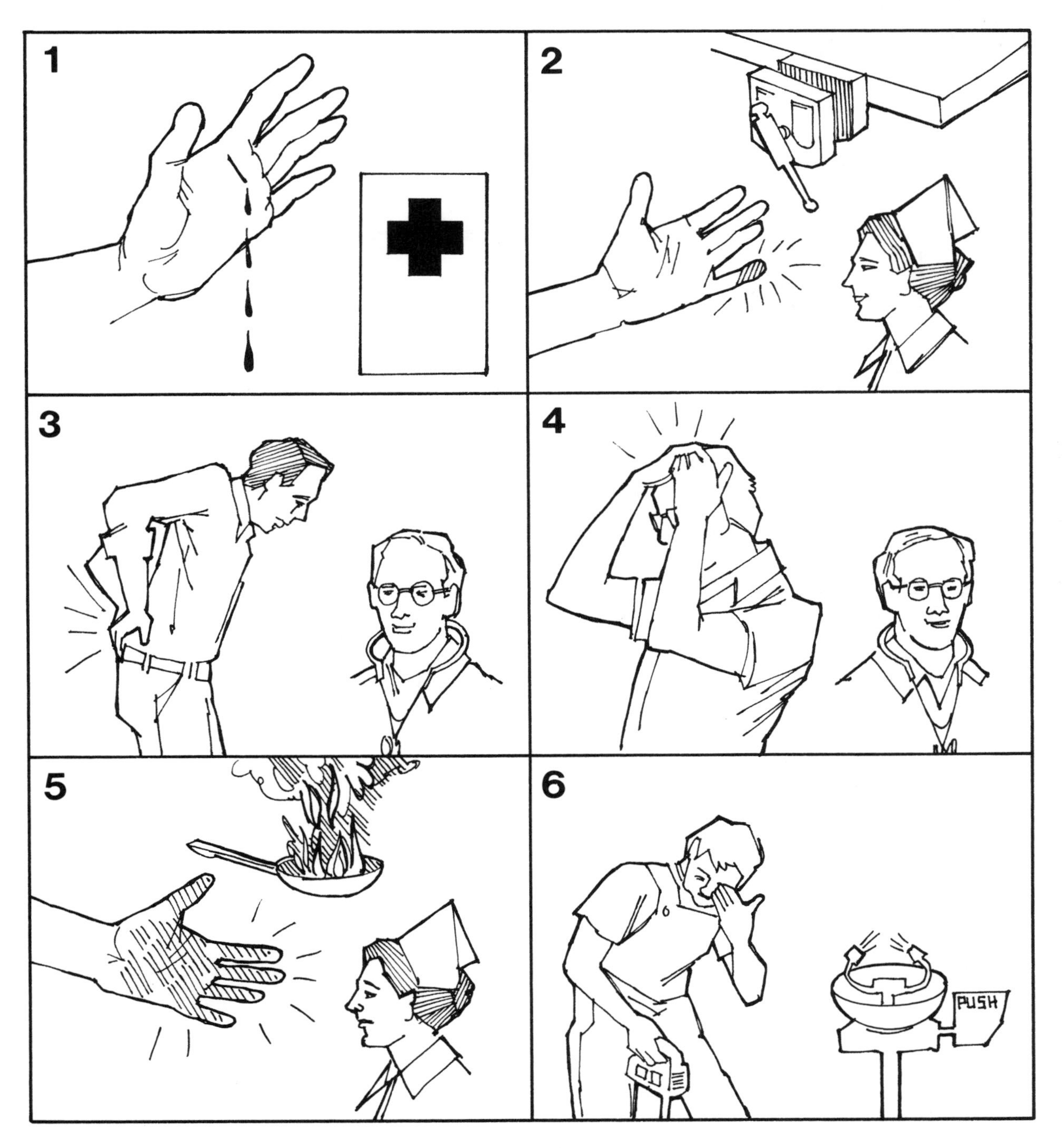

READ THE SAFETY POSTER.

Answer the question.

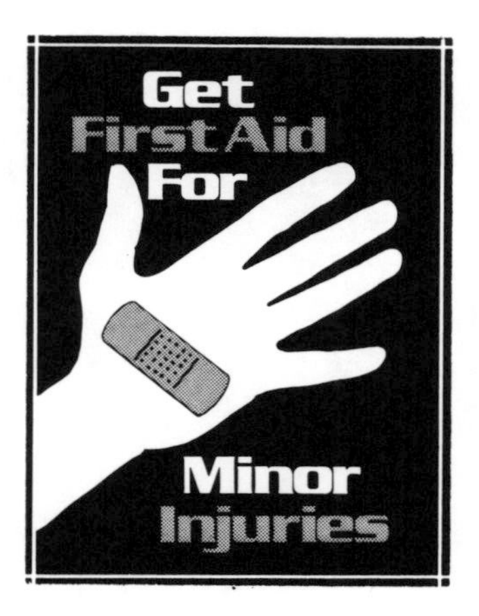

What should you do?

____

____

READ ABOUT SAFETY.

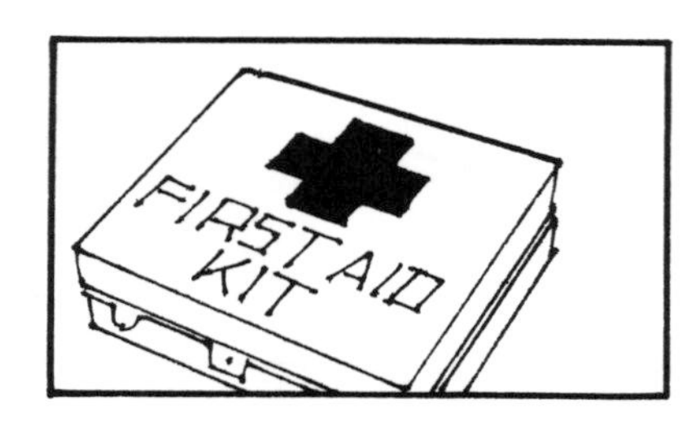

It is very important to take care of injuries immediately. If it is a minor injury, you can get first aid. All workplaces should have first aid kits. The kit has aspirin, bandages, disinfectant and other things.

Sometimes a worker gets something in his eye. Many workplaces have eye wash stations. A worker can wash his eyes.

Sometimes a workplace will have a nurse. He or she takes care of minor injuries. If there is a serious injury, a doctor comes or an ambulance takes the worker to the hospital.

CHECK YOUR UNDERSTANDING.

1. What kinds of medical assistance are available?

____

2. What happens if there is a serious injury?

____

ROLE-PLAY.

Read your role. Close your book. Role-play with your partner.

1. Two workers are talking.

A

You see an injured worker and ask what happened. You check if the injury is serious and give the worker advice.

B

You cut your hand on a piece of wire. It is a small cut.

2. A worker is talking to the boss.

A

You are the boss. You ask the worker what happened. You check if the injury is serious and give the worker advice.

B

You are the worker. You got something in your eye. It isn't serious.

3. Two workers are talking.

A

You see an injured co-worker and ask what happened. You check if the injury is serious and give the worker advice.

B

You got your finger caught in a machine. Your finger hurts a lot.

4. A worker is talking to the manager.

A

You are the manager. You see an injured worker and ask what happened. You give the worker advice.

B

You are the worker. You fell and hit your head. You don't feel well.

FOLLOW-UP.

Visit a workplace or shop class. What medical assistance is available?

LESSON SEVENTEEN: REPORTING AN INJURY

LISTEN.

Stefan's co-worker, Joe, reports the accident to the supervisor.

Joe:	I want to report an accident.
Supervisor:	What happened?
Joe:	Stefan cut his hand on a piece of metal. He went to get first aid.
Supervisor:	Is it serious?
Joe:	No, I don't think so.
Supervisor:	What happened?
Joe:	I'm not sure. I think he forgot to put his gloves on.
Supervisor:	I'll go see if he's OK. Can you take over his machine till he comes back?
Joe:	Sure, no problem.

CHECK YOUR UNDERSTANDING.

1. Joe is reporting Stefan's accident to the manager.	RIGHT	WRONG
2. Stefan went home.	RIGHT	WRONG
3. The supervisor is angry.	RIGHT	WRONG
4. Stefan lost his job.	RIGHT	WRONG

PRACTICE.

Look at the pictures. What happened?

Practice the conversation with your partner.
Practice conversations using the pictures.

Worker:	I want to report an injury.
Supervisor:	What happened?
Worker:	I sprained my ankle.
Supervisor:	How did it happen?
Worker:	I fell on the stairs.

Practice conversations using the pictures below.

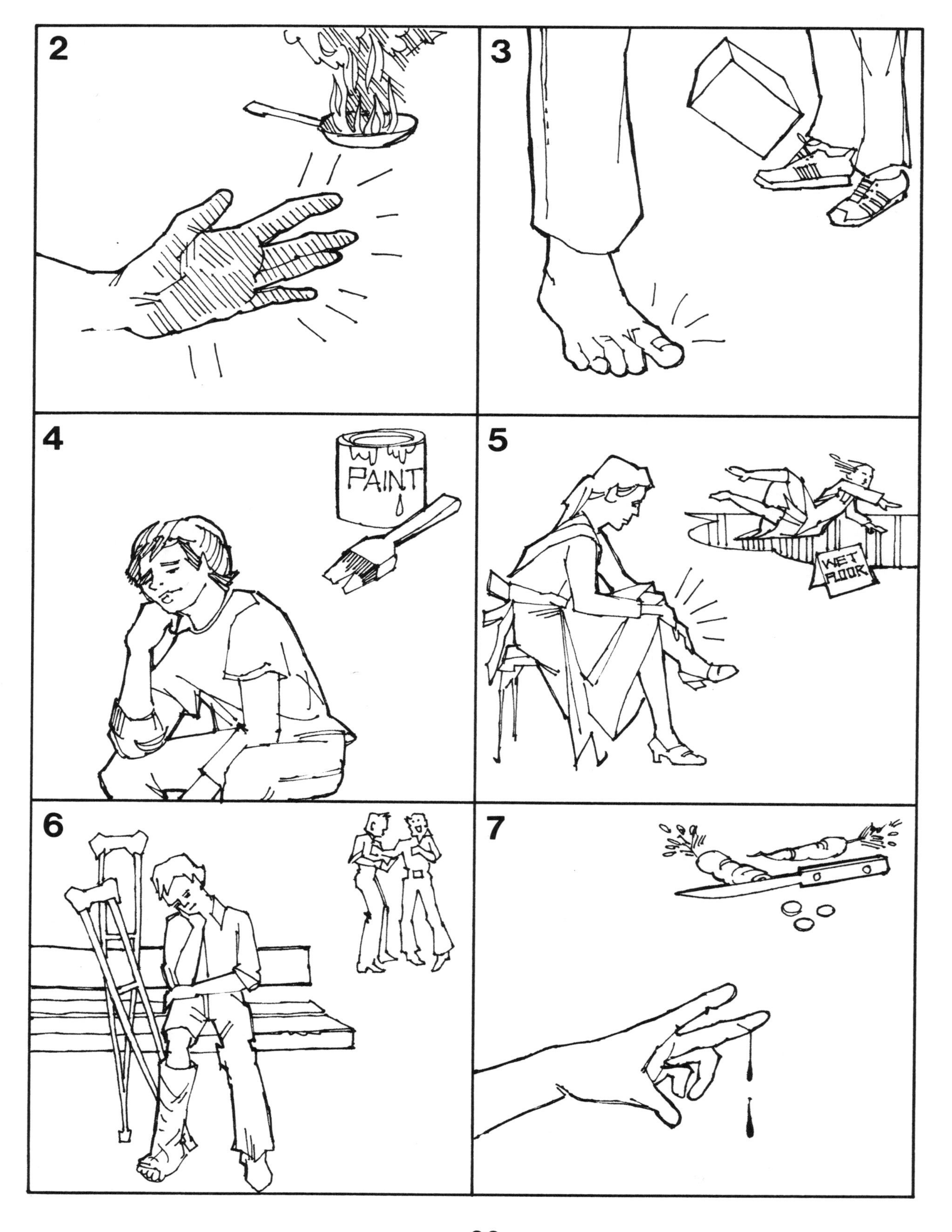

READ THE SAFETY POSTER.

Answer the question.

What should you do?

____

____

READ ABOUT SAFETY.

There is a law called the Workers' Compensation Act. Most workers are covered by this law. If a worker has a job-related injury or disease, he or she has benefits. The Act provides the worker with medical care. The injured worker also gets compensation for the days he or she cannot work.

It is very important that a worker report an injury to the employer as soon as possible. If the worker does not report the injury soon, he or she may not get the benefits.

Each state has different laws about the amount of benefits and the procedures for getting the benefits. It is important to find out about the benefits. A worker can ask the employer, the union, or the State Department of Labor.

CHECK YOUR UNDERSTANDING.

1. When should a worker report an injury to the employer?

   ____

2. Why is it important to report an injury immediately?

   ____

3. Does every state have the same Workers' Compensation benefits?

   ____

4. How can you find out about the Workers' Compensation benefits in your state?

   ____

FILL OUT THE FORMS.

Look at the pictures on pages 95 and 96. Report the accidents by filling out the forms. Use your name and today's date.

1 ACCIDENT REPORT FORM

NAME ____________________

DATE OF INJURY _____ TIME _____

TYPE OF INJURY	BODY PART INJURED	
____bruise	____ankle	____ear
____burn	____arm	____eye
____cut	____back	____leg
____fracture	____chest	____toe
____poisoning	____finger	
____shock	____foot	
____sprain	____hand	
____strain	____head	
OTHER: ____________	OTHER: ____________	

HOW DID THE ACCIDENT HAPPEN?

____

____

SIGNATURE ____________________

DATE ________

2 ACCIDENT REPORT FORM

NAME ____________________

DATE OF INJURY _____ TIME _____

TYPE OF INJURY	BODY PART INJURED	
____bruise	____ankle	____ear
____burn	____arm	____eye
____cut	____back	____leg
____fracture	____chest	____toe
____poisoning	____finger	
____shock	____foot	
____sprain	____hand	
____strain	____head	
OTHER: ____________	OTHER: ____________	

HOW DID THE ACCIDENT HAPPEN?

____

____

SIGNATURE ____________________

DATE ________

3 ACCIDENT REPORT FORM

NAME ____________________

DATE OF INJURY _____ TIME _____

TYPE OF INJURY	BODY PART INJURED	
____bruise	____ankle	____ear
____burn	____arm	____eye
____cut	____back	____leg
____fracture	____chest	____toe
____poisoning	____finger	
____shock	____foot	
____sprain	____hand	
____strain	____head	
OTHER: ____________	OTHER: ____________	

HOW DID THE ACCIDENT HAPPEN?

____

____

SIGNATURE ____________________

DATE ________

4 ACCIDENT REPORT FORM

NAME ____________________

DATE OF INJURY _____ TIME _____

TYPE OF INJURY	BODY PART INJURED	
____bruise	____ankle	____ear
____burn	____arm	____eye
____cut	____back	____leg
____fracture	____chest	____toe
____poisoning	____finger	
____shock	____foot	
____sprain	____hand	
____strain	____head	
OTHER: ____________	OTHER: ____________	

HOW DID THE ACCIDENT HAPPEN?

____

____

SIGNATURE ____________________

DATE ________

5 ACCIDENT REPORT FORM

NAME ______________________________

DATE OF INJURY ______ TIME ______

TYPE OF INJURY	BODY PART INJURED	
____bruise	____ankle	____ear
____burn	____arm	____eye
____cut	____back	____leg
____fracture	____chest	____toe
____poisoning	____finger	
____shock	____foot	
____sprain	____hand	
____strain	____head	
OTHER: ____________	OTHER: ____________	

HOW DID THE ACCIDENT HAPPEN?

____

____

SIGNATURE ______________________

DATE ________

6 ACCIDENT REPORT FORM

NAME ______________________________

DATE OF INJURY ______ TIME ______

TYPE OF INJURY	BODY PART INJURED	
____bruise	____ankle	____ear
____burn	____arm	____eye
____cut	____back	____leg
____fracture	____chest	____toe
____poisoning	____finger	
____shock	____foot	
____sprain	____hand	
____strain	____head	
OTHER: ____________	OTHER: ____________	

HOW DID THE ACCIDENT HAPPEN?

____

____

SIGNATURE ______________________

DATE ________

7 ACCIDENT REPORT FORM

NAME ______________________________

DATE OF INJURY ______ TIME ______

TYPE OF INJURY	BODY PART INJURED	
____bruise	____ankle	____ear
____burn	____arm	____eye
____cut	____back	____leg
____fracture	____chest	____toe
____poisoning	____finger	
____shock	____foot	
____sprain	____hand	
____strain	____head	
OTHER: ____________	OTHER: ____________	

HOW DID THE ACCIDENT HAPPEN?

____

____

SIGNATURE ______________________

DATE ________

ROLE-PLAY.

Read your role. Close your book. Role-play with your partner.

1. **A worker is talking to the boss.**

A

You are the boss. A worker has had an accident. You ask the worker what happened and ask if the injury is serious. You suggest that the worker see the nurse.

B

You are the worker. You fell on the stairs and hurt your back. You tell the boss about the accident.

2. **Two workers are talking.**

A

Your co-worker doesn't look well. You ask your co-worker what happened. You suggest that your co-worker report the accident to the supervisor and then see the nurse.

B

A box fell and hit you on the head. You have a bad headache.

3. **A shop student is talking to the shop teacher.**

A

You are the shop student. You cut your leg on a piece of metal. It is a very bad cut. You report the accident to the shop teacher.

B

You are the shop teacher. You ask the student what happened. You tell the student that you will get the nurse.

FOLLOW-UP.

Visit a workplace or shop class. Ask for an accident report form. Do you understand the form? Can you fill it out?

LESSON EIGHTEEN: HOW DID IT HAPPEN?

LISTEN.

Rosa had an accident. She is reporting it to the supervisor.

Rosa:	I want to report an accident. I fell on the stairs.
Supervisor:	Did you hurt yourself?
Rosa:	Yes. My back is a little sore.
Supervisor:	How did it happen?
Rosa:	Someone dropped a box of cookies on the stairs. I was walking down the stairs. I slipped and fell.
Supervisor:	I'll take a look at those stairs. They should be swept. Now maybe you'd better go see a doctor.
Rosa:	OK.

CHECK YOUR UNDERSTANDING.

1.	Rosa fell on the stairs.	RIGHT	WRONG
2.	The stairs were wet.	RIGHT	WRONG
3.	Rosa didn't get hurt.	RIGHT	WRONG
4.	The stairs are safe.	RIGHT	WRONG
5.	The supervisor is going to check the stairs.	RIGHT	WRONG

PRACTICE.

Look at the pictures. What happened?
Practice the conversation with your partner.
Practice conversations using the pictures below.

Worker:	I had an accident.
Supervisor:	What happened?
Worker:	I was driving the fork lift. The brakes were bad. I hit some boxes.
Supervisor:	Did you get hurt?
Worker:	Yes, I hit my head.
Supervisor:	You'd better see the nurse.
Worker:	OK.

PRACTICE.

Look at the pictures. What happened?
Practice conversations using the pictures below.

READ THE SAFETY POSTER.

Answer the question.

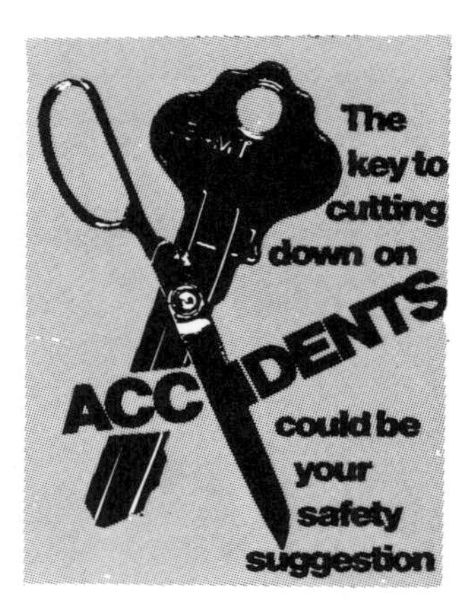

What should you do?

____

____

READ ABOUT SAFETY.

Each company has its own reporting forms and procedures. It is important to find out the procedures at the company you are working for. These are the rules for reporting accidents at Rosa's company:

IN CASE OF AN ACCIDENT

1. Report the accident to the supervisor immediately.
2. If the supervisor is not available, report the accident to personnel.
3. Report all accidents yourself.
4. A failure to report an accident immediately may results in the insurance company NOT paying any medical claims.
5. Fill out the accident report forms as soon as possible.

CHECK YOUR UNDERSTANDING.

1. Do all companies have the same reporting forms? __________
2. Who should Rosa report the accident to?

   ____

3. What could happen if the accident isn't reported immediately?

   ____

   ____

FILL OUT THE FORMS.

Look at the pictures on pages 102 and 103. Report the accidents by filling out the forms. Use your name and today's date and time.

1 ACCIDENT REPORT FORM

NAME: ________________________________

DATE OF ACCIDENT: ____________ TIME: ________

SIGNATURE: ____________________ DATE: ________

TYPE OF INJURY:

HOW DID THE ACCIDENT HAPPEN?

2 ACCIDENT REPORT FORM

NAME: ________________________________

DATE OF ACCIDENT: ____________ TIME: ________

SIGNATURE: ____________________ DATE: ________

TYPE OF INJURY:

HOW DID THE ACCIDENT HAPPEN?

3 ACCIDENT REPORT FORM

NAME: ______________________________

DATE OF ACCIDENT: __________ TIME: ________

SIGNATURE: ________________ DATE: ________

TYPE OF INJURY:

HOW DID THE ACCIDENT HAPPEN?

4 ACCIDENT REPORT FORM

NAME: ______________________________

DATE OF ACCIDENT: __________ TIME: ________

SIGNATURE: ________________ DATE: ________

TYPE OF INJURY:

HOW DID THE ACCIDENT HAPPEN?

5

ACCIDENT REPORT FORM

NAME: ______________________________

DATE OF ACCIDENT: ____________ TIME: ________

SIGNATURE: ____________________ DATE: ________

TYPE OF INJURY:

HOW DID THE ACCIDENT HAPPEN?

6

ACCIDENT REPORT FORM

NAME: ______________________________

DATE OF ACCIDENT: ____________ TIME: ________

SIGNATURE: ____________________ DATE: ________

TYPE OF INJURY:

HOW DID THE ACCIDENT HAPPEN?

ROLE-PLAY.

Read your role. Close your book. Role-play with your partner.

1. A worker is talking to the boss.

A

You are the worker. You want to report an injury. You were picking up a broken bottle and you cut yourself. It is not serious.

B

You are the boss. You ask the worker about the accident. You tell the worker that he or she should not pick up broken glass. Broken glass should be swept up with a broom.

2. A worker is talking to the supervisor.

A

You are the worker. You left your gloves at home. You cut your hands when you picked up a piece of sheet metal. You are reporting the injury to the supervisor.

B

You are the supervisor. You ask the worker about the injury. You remind the worker that he or she must always wear gloves. You are worried that the cuts will become infected. You take the worker to a doctor.

3. A worker is talking to the manager.

A

You are a machine operator. A sign on your machine said: DO NOT WEAR GLOVES WHILE OPERATING THIS MACHINE.
You did not see the sign.
You were wearing gloves. Your finger got caught in the machine.
You are reporting the accident to the manager.

B

You are the manager. You ask the worker about the accident and you remind the worker to read the safety signs. You take the worker to the hospital.

FOLLOW-UP.

Visit a workplace or shop class. What are common accidents and injuries? What causes the accidents?

LESSON NINETEEN: MORE SUGGESTIONS

LISTEN.

Champa had an accident. She is talking to her supervisor.

Champa:	Excuse me, I'd like to report a minor accident.
Supervisor:	What happened?
Champa:	I was walking down the aisle and I tripped on a box.
Supervisor:	Are you hurt?
Champa:	My toe is a little sore, but it's all right. Somebody could really get hurt, though. There shouldn't be any boxes in the aisle. It's dangerous. I think you should check it out.
Supervisor:	Why are the boxes in the aisle?
Champa:	I think there are too many supplies on the shop floor. They should be kept in the store room until they are needed.
Supervisor:	You could be right. I'll check it out. Thanks for the suggestion.

CHECK YOUR UNDERSTANDING.

1. Champa had a bad accident.	RIGHT	WRONG
2. Champa broke her leg.	RIGHT	WRONG
3. Boxes shouldn't be in the aisle.	RIGHT	WRONG
4. The supervisor is happy to get Champa's suggestion.	RIGHT	WRONG

PRACTICE.

Look at the pictures. What is unsafe? Circle the problem.

Practice the conversation with your partner.
Practice conversations using the pictures below.

Worker:	Some machine guards are broken. It's dangerous. I think they should be fixed immediately.
Supervisor:	Thanks for the suggestion. I'll check it out right away.

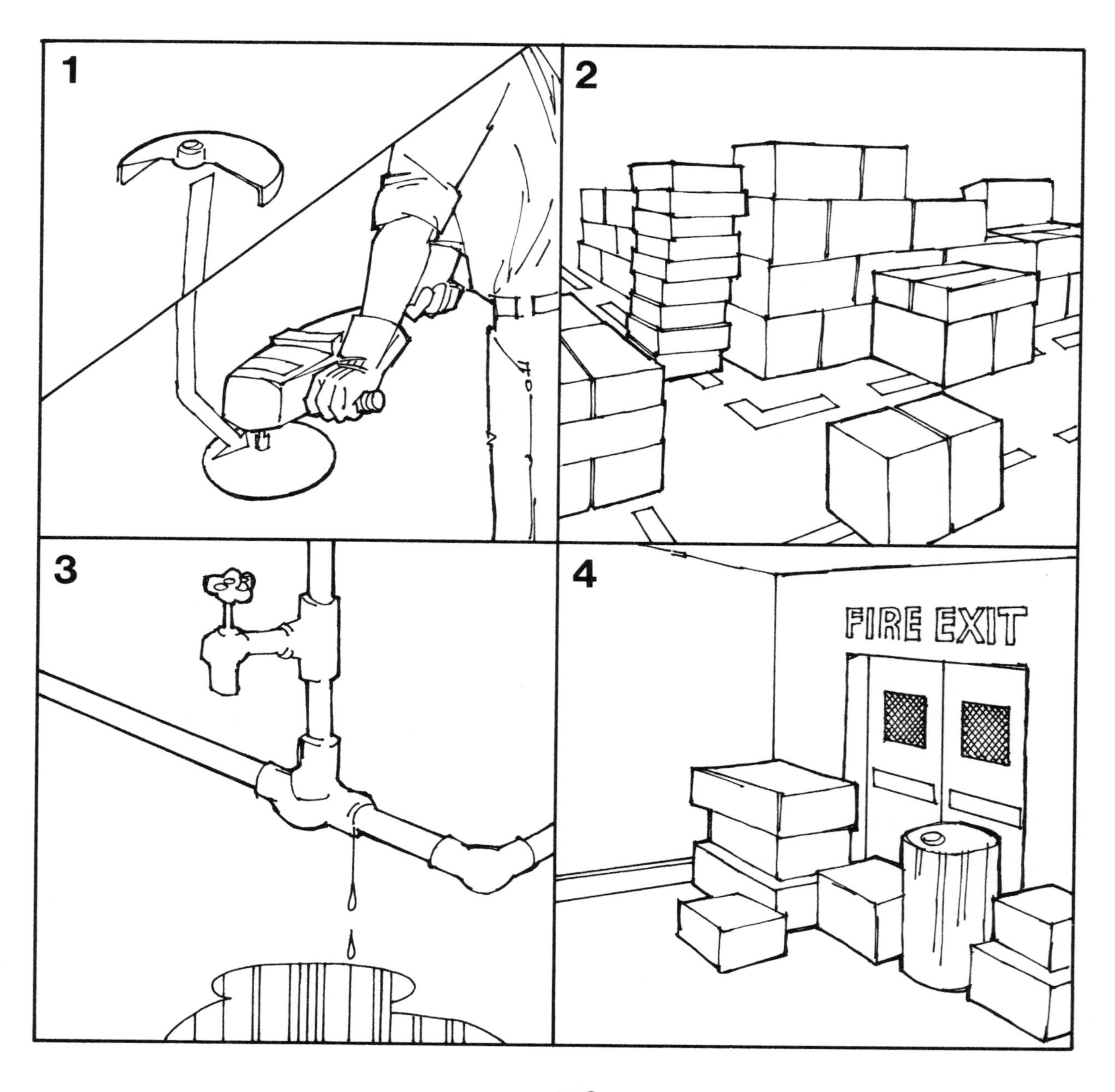

READ THE SAFETY POSTER.

Answer the question.

What should you do?

____

____

READ ABOUT SAFETY.

You should report every accident. When you report an accident, you will probably be asked these questions:

1. Were you injured?
2. How did the accident happen?
3. How can another accident be prevented?

It is very important to make suggestions to prevent other accidents from happening!

CHECK YOUR UNDERSTANDING.

1. When you report an accident, what will people ask you?

____

____

____

2. Why is it important to make safety suggestions?

____

____

____

FILL OUT THE FORMS.

Look at the pictures. Each person had an accident. What caused the accident? What suggestions can you make to prevent another accident?

Fill out a form for each picture. Use your name and today's date and time.

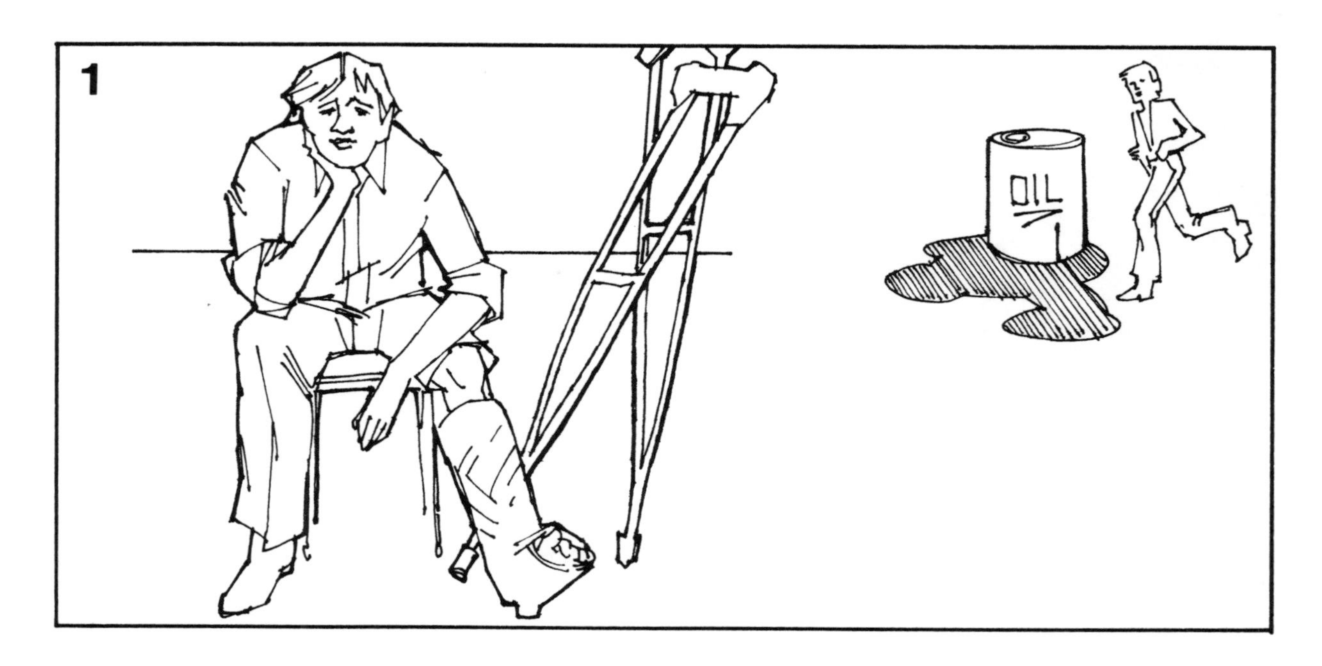

ACCIDENT REPORT FORM
NAME ______________ DATE ______________ DATE OF INJURY ______________ TIME ________
DESCRIBE THE INJURY:
DO YOU HAVE A SUGGESTION FOR PREVENTING THIS TYPE OF ACCIDENT?
DESCRIBE HOW THE ACCIDENT HAPPENED:
SIGNATURE ____________________ DATE __________

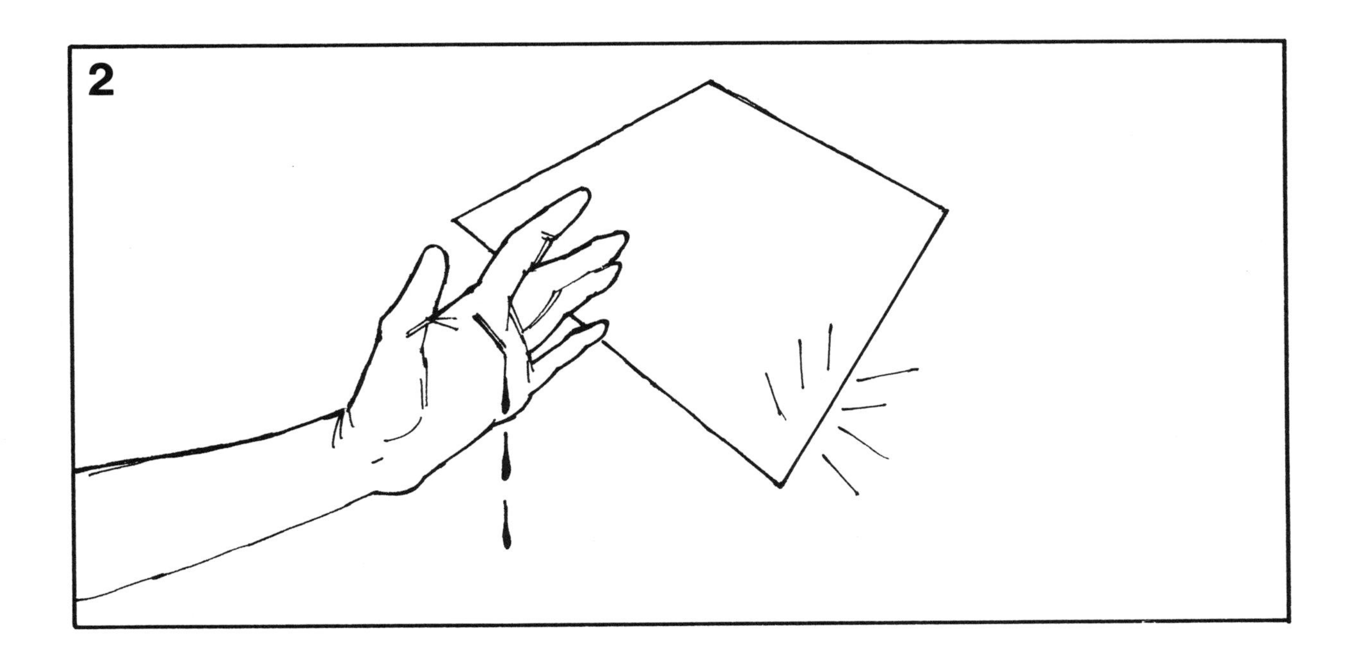

ACCIDENT REPORT FORM

NAME ________________ DATE ________________

DATE OF INJURY ________________ TIME ________

DESCRIBE THE INJURY:

DO YOU HAVE A SUGGESTION FOR PREVENTING THIS TYPE OF ACCIDENT?

DESCRIBE HOW THE ACCIDENT HAPPENED:

SIGNATURE ______________________________

DATE ____________

ROLE-PLAY.

Read your role. Close your book. Role-play with your partner.

1. A student is talking to the shop teacher.

A

You are the student. You tripped on some tools which were on the floor. You did not hurt yourself but you think a serious accident might happen. You make a suggestion to the shop teacher.

B

You are the shop teacher. You listen to the student and thank him or her for the suggestion.

2. A worker is talking to the supervisor.

A

You are the worker. The floor was wet and you fell. You did not hurt yourself but you are worried about other people getting hurt. You make a suggestion about preventing falls.

B

You are the supervisor. You listen to the worker's suggestion and thank him or her for the suggestion.

3. A worker is talking to personnel.

A

You are the worker. You were in a hurry so you did not use the machine guard. Your hand got caught in the machine. You broke a finger. You make a suggestion about preventing this kind of accident.

B

You are personnel. You are very interested in preventing accidents and injuries. You ask the worker about the accident. Thank the worker for the suggestion.

FOLLOW-UP.

Visit a workplace or shop class. What kinds of accidents and injuries happen?
What kinds of suggestions have been made?
Can you think of any more suggestions?